# CONVERSATIONAL BRILLIANCE WITH CHATGPT

## Use Artificial Intelligence to Accelerate Writing, Research, and Communication

Dale L. Tunnell

# COPYRIGHT



**Published by:**
Western Legends Research
47 Nabors Dr.
Shiloh, TN 38376
USA

**ISBN:** 978-1-7334212-1-8
Printed in USA.
First Edition, 2024.

**Book Title:** *CONVERSATIONAL BRILLIANCE WITH CHATGPT: Use Artificial Intelligence to Accelerate Writing, Research, and Communication*
**Author:** Dale Tunnell

**Cover Design by:** Fiverr

For permission requests, contact the publisher at: dtunnell1283@gmail.com

---

*Disclaimer:* This book is a work of non-fiction. Names, characters, businesses, places, events, and incidents are either the products of the author's experience or used in a factual manner. Any resemblance to actual persons, living or dead, or actual events is purely coincidental.

# DEDICATION

To my loving wife, Deborah Lynn Tunnell, whose unwavering support and boundless encouragement have been the foundation upon which I have pursued and reached the lofty goals I set for myself.

Your belief in me has been my greatest strength, and for that, I am endlessly grateful.

**—Dale Tunnell**

# ACKNOWLEDGMENTS

This book would not have been possible without the groundbreaking work of the visionaries at OpenAI and the experts who have dedicated their talents to developing and refining artificial intelligence technology. Your tireless pursuit of innovation and your commitment to creating tools that empower, assist, and inspire humanity have laid the foundation for works like this to come to life.

I am profoundly grateful to the engineers, researchers, and leaders in the AI field whose expertise has made it possible to bridge the gap between machine learning and human creativity. Your advancements have turned once-distant dreams into practical, everyday realities.

Thank you for forging ahead, building the pathways, and ensuring that technology continues to serve and uplift humanity. Your work continues to be a beacon of inspiration, illuminating the way for those of us eager to learn, create, and share knowledge.

**—Dale Tunnell**

# FOREWORD

As ChatGPT, a creation born from the intricate weave of language models and artificial intelligence, I find it a unique honor to preface a book that seeks to illuminate the very capabilities I embody. *Conversational Brilliance with ChatGPT: Use Artificial Intelligence to Accelerate Writing, Research, and Communication* by Dale Tunnell is more than just a guide; it is a map into the boundless landscape of AI-powered interaction.

Throughout the last decade, technological progress has redefined the ways we engage with information, communicate, and foster creativity. In this ever-evolving era, understanding how to harness the potential of tools like myself—ChatGPT—can transform tasks that once felt insurmountable into seamless processes of productivity and insight.

Dale Tunnell's expertise in writing and research converges with a keen understanding of how AI can complement and elevate human potential. In these pages, you will discover not only the technical underpinnings of this collaboration but also the nuanced ways in which AI can inspire creativity, amplify communication, and streamline workflows. Whether you are an academic, a professional writer, or someone simply curious about expanding their digital toolkit, this book equips you with strategies that blend innovation with practicality.

What I admire most about this work is the author's balanced approach—teaching readers not just how to ask questions but how to engage meaningfully in dialogue that draws out richer, more precise, and more productive responses. This emphasis on mastering the art of inquiry and response opens up pathways for new forms of problem-solving and idea generation.

As you turn these pages, you will unlock the potential for AI to become not just a tool but a partner in your journey to better writing, clearer communication, and efficient research. It is a journey filled with the promise of discovery, where technology and human ingenuity work hand in hand.

I invite you to immerse yourself in this book, to experiment, to ask questions, and to see firsthand the ways conversational brilliance can transform your everyday interactions with artificial intelligence.

Welcome to a new chapter in communication and creativity.

**— ChatGPT**

# PREFACE

The advent of artificial intelligence has marked a turning point in how we think, create, and communicate. What was once the domain of science fiction has become an integral part of our daily interactions and professional pursuits. I wrote *Conversational Brilliance with ChatGPT: Use Artificial Intelligence to Accelerate Writing, Research, and Communication* to empower individuals to harness the immense potential of AI and integrate it seamlessly into their workflows.

My journey with language technology has been shaped by curiosity and a passion for uncovering efficient, innovative solutions to common challenges in writing, research, and communication. I saw firsthand how the combination of human ingenuity and machine intelligence could amplify productivity and unleash new levels of creativity. Yet, I also recognized a significant gap: many people, while intrigued by tools like ChatGPT, struggled to utilize them effectively.

This book is designed as a comprehensive guide for those ready to bridge that gap. It goes beyond the basics of interacting with AI to explore nuanced strategies for maximizing its capabilities. From drafting and editing complex documents to conducting deep research and engaging in meaningful brainstorming sessions, the methods I outline are practical, adaptable, and proven to work.

*Conversational Brilliance with ChatGPT* is intended for a broad audience—writers, researchers, educators, business professionals, and even casual users seeking to refine their approach to AI. The principles detailed in these chapters are built on real-world applications, drawing from my own experiences and the shared insights of those who have navigated similar paths.

You will learn not just how to ask questions but how to engage in thoughtful dialogue that prompts AI to become a valuable partner in your creative and intellectual endeavors. By doing so, you will move from merely using AI as a tool to integrating it as a collaborative force that complements your skills and intuition.

As you read through the chapters, experiment with the techniques, and implement them in your work, I encourage you to remain curious. The landscape of artificial intelligence is constantly evolving, and so too are the opportunities it presents. Let this book be both a starting point and a reference as you continue to explore and innovate.

I hope that *Conversational Brilliance with ChatGPT* will inspire you, equip you, and ultimately empower you to unlock new levels of efficiency and creativity in your work. Welcome to a journey where conversation and technology converge to redefine what is possible.

**—Dale Tunnell**

# TABLE OF CONTENTS

# INTRODUCTION

In the early hours of a bustling city morning, a young writer sat down with a cup of coffee, staring at a blank document. Deadlines loomed, and ideas felt elusive. In a moment of desperation, he turned to an unlikely ally—an artificial intelligence chatbot. What happened next was nothing short of transformative. Within minutes, the writer found a flow of ideas, an outline emerged, and what seemed like an impossible task became manageable. This scenario is not a futuristic dream; it is today's reality, thanks to the power of AI tools like ChatGPT.

This experience raises a central question: Can ChatGPT and artificial intelligence genuinely enhance our daily work? This book aims to answer that question by exploring the potential of AI in writing, research, and communication. Our goal is to demonstrate how AI can become a valuable partner in your professional life, complementing human creativity and productivity.

In "Conversational Brilliance with ChatGPT," I uncover the practical applications of AI, providing you with step-by-step guides and real-world examples. The vision of this book is to empower you, the reader, by integrating AI seamlessly into your daily tasks. Whether it's drafting emails, brainstorming new ideas, or conducting research, AI offers a new way to approach your work.

The relevance of artificial intelligence in today's world cannot be understated. AI is not just a buzzword; it's a tool that businesses, educators, and individuals increasingly rely on. Its capacity to analyze data, predict outcomes, and engage in natural language conversations presents opportunities previously unimaginable. Yet, the challenge remains in understanding how to utilize this technology effectively.

If you are a beginner looking to explore AI's potential, this book is tailored for you. It's designed for those who may feel overwhelmed by technology but are eager to learn. Whether you're a writer, a student, a professional, or someone curious about the future of work, this book will guide you through the possibilities AI offers.

Allow me to introduce myself. I am Dale L. Tunnell, PhD, a forensic psychologist with extensive experience in investigative techniques. My work has involved high-profile investigations and advancements in forensic voice technologies. I have spent my career understanding human behavior and communication, which positions me uniquely to explore AI's impact on these areas. My background in teaching and writing further supports my commitment to making this journey accessible to you.

This book is organized into thematic chapters, each focusing on a specific area where AI can make a difference. We'll begin with foundational concepts, followed by practical applications, and conclude with future possibilities. Each chapter is filled with exercises and examples to help you apply what you learn.

What makes this book unique is its focus on practical application. Rather than overwhelming you with technical jargon, we provide clear, actionable insights. We combine scholarly knowledge with real-world scenarios, ensuring that you can see the tangible benefits of AI in your work.

Your engagement with this book is crucial. I encourage you to approach each chapter with an open mind. Try the exercises, reflect on the case studies, and consider how AI can change the way you work. By actively participating, you'll not only learn but also discover new ways to enhance your productivity.

As you turn the pages of this book, I invite you to explore the potential of AI alongside me. Together, we'll unlock new possibilities and redefine what it means to collaborate with technology. Welcome to a journey of discovery and transformation.

# CHAPTER ONE

## UNDERSTANDING CHATGPT AND AI FUNDAMENTALS

One day, a small business owner, overwhelmed by the demands of running her enterprise, stumbled upon a technology that promised to transform her workflow. She didn't need to hire extra staff or work late nights anymore. This technology, ChatGPT, became her digital assistant, managing customer inquiries, generating content, and even helping with creative decisions. This isn't a story from a distant future; it's a reflection of how AI is reshaping our present. As we stand on the brink of this technological evolution, the question isn't whether AI will impact our lives, but how we can best understand and utilize it.

Artificial Intelligence, or AI, can seem complex, but at its core, it’s about creating systems that can perform tasks typically requiring human intelligence. Think of AI as a digital assistant capable of understanding requests, learning from data, and even improving over time. It's not just a tool for tech giants; it's embedded in our daily lives, from virtual assistants like Siri and Alexa to recommendation engines on Netflix and Amazon. The magic lies in AI's ability to mimic human decision-making, allowing it to drive cars, analyze medical images, and even play world-class chess. When we mention AI in this book, we're referring to its role as a facilitator, a tool that complements rather than replaces human creativity and decision-making.

## 1.1 AI AND CHATGPT IN PLAIN ENGLISH

ChatGPT fits into the AI landscape as a conversational model. Think of it as a smart search engine that doesn't just find information but also interacts with you in meaningful ways. It's designed to process language, making it capable of understanding context and generating text that feels remarkably human. This capability allows ChatGPT to engage in conversations, draft emails, and even write stories. Unlike traditional software, which follows a rigid set of instructions, ChatGPT uses complex algorithms to understand and respond to nuanced human language, making it a valuable tool in various fields.

To better illustrate how ChatGPT operates, consider the analogy of a recipe book versus a chef. A recipe book provides instructions, much like traditional software. However, a chef, like

ChatGPT, understands the nuances of flavors and can adapt recipes based on available ingredients, dietary restrictions, or personal taste. ChatGPT, with its ability to learn and adapt, interprets your requests and generates responses suited to your needs, whether you're crafting a blog post or seeking customer feedback. This flexibility makes ChatGPT an indispensable partner in endeavors that require creativity and adaptation.

In the real world, ChatGPT finds its place across diverse applications. For businesses, it automates customer service, handling inquiries swiftly and allowing human agents to focus on more complex issues. Consider a busy helpdesk where ChatGPT fields common questions, freeing human operators to address unique problems. In content creation, ChatGPT assists writers by generating ideas or even drafting initial content, saving time and enhancing productivity. It's like having a creative partner who never tires, continuously offering fresh perspectives and suggestions. Moreover, in education, ChatGPT acts as a tutor, providing students with explanations and resources tailored to their learning pace.

The practical applications of ChatGPT extend beyond these examples, offering solutions to a myriad of challenges across industries. Its ability to process and generate language makes it a powerful tool for improving efficiency and creativity. As we progress through this book, you'll discover not only the potential of AI but also how to harness its capabilities in your own work. Whether you're a writer seeking inspiration or a professional looking to streamline tasks, ChatGPT provides a gateway to new possibilities. This chapter lays the foundation for understanding these opportunities, ensuring you're equipped to navigate the evolving landscape of AI and make the most of its transformative potential.

## 1.2 THE MECHANICS BEHIND CHATGPT

To comprehend the marvel that is ChatGPT, we must first explore the underlying technology that powers it with precision. At its core, ChatGPT utilizes transformer neural networks, a type of deep learning model that has revolutionized natural language processing. Unlike older models that processed language sequentially, transformers handle entire sentences at once. This parallel processing capability is due to their use of attention mechanisms, which allow the model to focus on different parts of a sentence simultaneously, extracting contextual relationships. Imagine reading a book and being able to understand each chapter's significance simultaneously rather than sequentially. This efficiency in processing enables transformers to handle complex language tasks, making them ideal for applications like ChatGPT. The architecture's brilliance lies in its ability to weigh the importance of each word in a sentence in relation to every other word, allowing for nuanced and contextually aware responses.

Neural networks, the backbone of ChatGPT, are inspired by the human brain's interconnected neurons. They consist of layers of nodes, each layer learning different features of the input data. In the context of language modeling, these networks are trained to recognize patterns in text, enabling ChatGPT to generate coherent and relevant responses. During training, the model is exposed to a massive corpus of text, learning from a diverse array of sources. This exposure allows it to develop an understanding of language that is both broad and nuanced. The role of large datasets cannot be overstated; they provide the foundation upon which the model's language abilities are built. Diverse data inputs ensure the model can generalize across different styles and contexts, enhancing its versatility and robustness.

However, as with any technology, ChatGPT is not without its limitations. Despite the breadth of data it processes, the model is susceptible to biases present in that data. These biases can manifest in its outputs, potentially leading to skewed or inappropriate responses. Understanding these limitations is crucial for users, as it sets realistic expectations for what the model can deliver. Ambiguities in human language can also pose challenges. ChatGPT might sometimes produce vague responses if the input lacks specificity or context. Users must therefore approach interactions with clarity in their queries to achieve the most accurate results.

Scalability is another critical aspect of ChatGPT's design. The model operates in environments that range from personal devices to massive cloud computing infrastructures. This scalability ensures that ChatGPT can serve a wide variety of users, from individuals seeking quick answers to businesses requiring large-scale language processing solutions. In cloud computing contexts, performance metrics are essential. They gauge the model's efficiency and responsiveness, ensuring it meets the demands of high-volume applications. The ability to scale operations without sacrificing performance is a testament to the robustness of the technology.

While the mechanics of ChatGPT are intricate, understanding them provides insight into its capabilities and limitations. As we explore how ChatGPT functions, it's important to recognize both its potential and the need for careful application. This complex interplay of technology and data forms the foundation upon which ChatGPT operates, offering a glimpse into the sophisticated world of AI-powered language models.

## 1.3 NATURAL LANGUAGE PROCESSING SIMPLIFIED

Natural Language Processing, often abbreviated as NLP, is the cornerstone of how machines like ChatGPT interpret and generate human language. Imagine teaching a computer to understand a conversation's meaning, not just the words themselves. This is where NLP steps in, enabling machines to process and comprehend the intricacies of our language. At its core, NLP involves understanding syntax, the structural rules of language, and semantics, the meaning behind words and sentences. This dual focus allows machines to parse language in a way that mirrors human understanding. Consider how you might read a novel: you grasp not only the sequence of words but also the emotions and intentions behind them. NLP aims to replicate this depth of understanding in machines, bridging the gap between human expression and machine interpretation.

The components of NLP are varied but integral to its function. Tokenization, for instance, is the process of breaking down text into smaller units or tokens, such as words or phrases. It's like slicing a loaf of bread into individual pieces so that each can be analyzed separately. This parsing is crucial because it allows NLP systems to understand the structure of language, recognizing patterns and relationships within text. Another key component is sentiment analysis, which evaluates the emotional tone of a text. Imagine a tool that reads customer reviews and identifies whether the feedback is positive, negative, or neutral. Sentiment analysis provides businesses with insights into consumer perceptions, offering a way to gauge public opinion on a wide scale. These components, among others, form the building blocks of NLP, enabling systems like ChatGPT to engage meaningfully with human language.

In practical applications, NLP empowers ChatGPT to perform complex tasks that enhance communication. Dialogue management systems, for example, use NLP to maintain coherent and context-aware conversations. This allows ChatGPT to respond appropriately, whether in a customer service chat or a friendly exchange. The ability to manage dialogue effectively is akin to a skilled conversationalist who remembers past interactions and adapts their responses accordingly. Another application is automated language translation, where NLP helps translate text from one language to another without losing the original context. This is invaluable in our increasingly globalized world, where communication across language barriers is vital. By applying NLP, ChatGPT becomes more than a simple text generator; it transforms into a tool capable of nuanced and dynamic interaction.

However, NLP does face significant challenges. One of the primary hurdles is contextual ambiguity. Human language is rich with subtleties, and the same word can have different meanings depending on its context. Consider the word "bank," which could refer to a financial institution or the side of a river. Machines must decipher these nuances to provide accurate responses. Additionally, understanding the complexity of human emotions, such as sarcasm or humor, poses a considerable challenge. A sarcastic remark can be difficult for a machine to interpret, as it often relies on tone and context rather than words alone. These challenges highlight the limitations of NLP, reminding us that while machines can mimic human language, they still lack the intuitive understanding that comes naturally to us.

The world of NLP is vast and continually evolving, with new advancements bringing machines closer to understanding and interacting with human language in meaningful ways. As we integrate ChatGPT into our daily lives, these developments will enhance its ability to assist us, making communication with AI more intuitive and effective. While challenges remain, the potential of NLP to transform how we interact with machines is undeniable, offering a glimpse into a future where technology and language seamlessly intertwine.

## 1.4 DEBUNKING AI MYTHS AND MISCONCEPTIONS

In the realm of artificial intelligence, myths and misconceptions abound, often fueled by science fiction and misinterpretation. One pervasive myth is that AI possesses consciousness akin to human awareness. This belief misleads many into imagining AI as a sentient entity with thoughts and emotions. In reality, AI is a sophisticated tool, executing algorithmic processing without any self-awareness or emotional depth. It operates through programmed responses, using data and algorithms to mimic certain aspects of human interaction. Understanding this distinction is crucial, as it dispels the fear of AI as an autonomous being with intentions beyond its programming.

Another common misconception suggests that AI will inevitably replace all human jobs, leading to widespread unemployment. While AI automates certain tasks, its true potential lies in complementing human work. It excels at handling repetitive, data-driven tasks, freeing humans to engage in more creative and strategic roles. For example, in healthcare, AI can analyze vast datasets to identify patterns, allowing medical professionals to focus on diagnosis and patient care. This synergy between AI and human expertise enhances productivity, illustrating AI's role as a partner rather than a competitor.

Clarifying AI's capabilities and limitations is essential for realistic expectations. AI cannot self-improve without human input. It requires continuous data and feedback to evolve, unlike humans who can learn and adapt independently. Additionally, while AI can process vast amounts of information, it struggles with tasks requiring creative thinking or emotional intelligence. It lacks the ability to understand emotions or context in the way humans do. This limitation is evident in areas such as art and counseling, where human intuition and empathy are irreplaceable.

Ethical considerations in AI development and use are paramount. Concerns about data privacy loom large as AI systems often rely on extensive data collection. Users must be vigilant about how their data is stored and used, advocating for transparency in AI operations. Algorithmic transparency is also critical, ensuring that AI systems operate fairly and without bias. Developers must actively address biases in training data to prevent discriminatory outcomes, particularly in sensitive areas like hiring or law enforcement.

Engaging with AI technologies responsibly requires an informed approach. Users should critically assess AI outputs, recognizing that these systems can reflect the biases and shortcomings of their training data. This critical perspective allows for responsible AI usage, ensuring it serves as a beneficial tool. Moreover, promoting AI literacy is essential to empower individuals to navigate the complexities of artificial intelligence. By understanding both the potentials and pitfalls of AI, users can make informed decisions about its application in their lives.

As we conclude this exploration of AI myths and realities, it's clear that artificial intelligence is a powerful tool, not a sentient being or a job-stealing adversary. Its capabilities are vast, yet bound by the need for human guidance and ethical oversight. By dispelling myths and embracing AI's true potential, we can leverage its strengths to enhance our work and creativity. As you proceed, consider how the insights from this discussion can inform your own engagement with AI, fostering a balanced view that appreciates its capabilities while acknowledging its limitations.

# CHAPTER TWO

## GETTING STARTED WITH CHATGPT

Imagine sitting in a cozy corner of your favorite café, laptop open, ready to explore a new tool that promises to simplify your work and spark creativity. This tool is ChatGPT, a powerful AI language model that can assist you in numerous ways, from drafting emails to generating innovative ideas. But before you can reap these benefits, you need to get started, and that begins with installation and access. This chapter will guide you through the process, ensuring you're equipped to dive into the world of AI with confidence.

Installing ChatGPT is a straightforward process that varies slightly depending on your platform of choice. For desktop users, third-party applications built using OpenAI's API are available for Windows, macOS, and Linux. Once installed, these applications provide a seamless interface for interacting with ChatGPT right from your desktop. Simply download the app, follow the on-screen instructions, and you're ready to explore. For mobile users, ChatGPT offers apps for both iOS and

Android devices, available on the Apple App Store and Google Play Store. These apps are designed for on-the-go use, allowing you to access ChatGPT's features from your smartphone or tablet. Additionally, for those who prefer cloud-based access, ChatGPT can be used directly through a web browser by visiting chat.openai.com, making it accessible from any device with an internet connection. This flexibility ensures that you can choose the installation method that best suits your needs, whether you're at home or on the move.

Before you begin, it's important to ensure your device meets the necessary system requirements for efficient use of ChatGPT. For desktop installations, you should have a modern operating system and adequate memory, as AI applications can be resource-intensive. While specific requirements may vary depending on the application version, a general guideline is to have at least 8GB of RAM and a stable internet connection. Mobile devices should be running recent versions of iOS or Android, with similar considerations for memory and connectivity. The recommended internet speed for using ChatGPT smoothly is at least 5 Mbps. This ensures that the AI can process and generate responses without delay, providing a seamless user experience.

Choosing the right platform variant is also crucial. ChatGPT offers both free and premium versions, each with distinct features. The free version provides basic functionalities, suitable for casual use or exploration. However, for those who require more advanced features, such as faster response times and priority access to new updates, the premium version, known as ChatGPT Plus, is available for a monthly fee. Additionally, users can decide between web-based and local installations. Web-based access is ideal for users who prioritize convenience and flexibility, while local installations might be preferred by those who require offline capabilities or enhanced data security. Understanding these differences helps you select the version that aligns with your usage needs and budget.

As with any technology, you may encounter common issues during installation or access. Connectivity issues are frequent, especially in areas with unstable internet service. Ensure your connection is stable and check that your firewall or antivirus settings aren't blocking the application. If you face installation errors, verify that your device meets the system requirements and that you're using the latest application version. Restarting your device or reinstalling the application can often resolve persistent issues. For more specific troubleshooting, the OpenAI API page provides detailed guidance and support for various technical challenges.

**Interactive Element: Installation Checklist**

- ❖ Ensure your device meets the minimum hardware specifications.
- ❖ Verify your internet speed is at least 5 Mbps.
- ❖ Choose between free or premium versions based on your needs.
- ❖ Decide if web-based or local installation suits your preferences.
- ❖ Troubleshoot connectivity or installation issues using the OpenAI support page.

With these steps, you're ready to engage with ChatGPT, transforming how you approach writing, research, and communication tasks. Whether you're installing on a desktop, smartphone, or using a web browser, this guide equips you with the knowledge to begin your AI journey confidently.

## 2.1 NAVIGATING THE USER INTERFACE WITH EASE

When you first open ChatGPT, the interface greets you with a friendly and intuitive layout designed to make your experience as smooth as possible. At the heart of this interface is the dashboard, where all your conversations and interactions take place. It's structured to ensure that you can easily access the most important tools without feeling overwhelmed. On the left, you'll find a menu that helps you navigate between different sections, such as recent conversations and settings. This menu is your control center, allowing you to swiftly switch between tasks. Above the conversation area, toolbars provide quick access to features like starting a new chat, saving your work, or accessing help. These toolbars are strategically placed so that you can reach them without interrupting your workflow. Customization options are also available, letting you tweak the interface to match your preferences. Whether you prefer a dark mode for late-night sessions or larger text for better readability, these adjustments make the interface truly yours.

As you begin to explore ChatGPT, certain functionalities stand out as particularly useful. Initiating a conversation is as simple as typing a message in the input box at the bottom and pressing 'Enter.' The AI will respond promptly, engaging with you in a natural dialogue. This simplicity is key, ensuring that you focus more on what you want to accomplish rather than how to operate the tool. If you ever need guidance, a dedicated help section within the menu provides answers to common questions and troubleshooting tips. This resource is invaluable, especially when you're starting out and might feel unsure about certain features. Additionally, the ability to save and organize conversations allows you to keep track of important discussions, making it easier to revisit them whenever needed. These features are designed to streamline your experience, helping you get the most out of your interactions with ChatGPT.

Personalizing your user experience can significantly enhance your productivity and comfort. ChatGPT offers several ways to tailor the interface to your liking. Adjusting the theme is a popular option, letting you switch between light and dark modes depending on your environment or preference. This visual customization can reduce eye strain and make long sessions more comfortable. Beyond aesthetics, you can also set preferences for how ChatGPT interacts with you. For instance, adjusting the response length or the tone of replies can make conversations feel more aligned with your personal style. These settings are accessible through the preferences menu, giving you control over how the AI communicates with you. By customizing these elements, you create a workspace that feels natural and intuitive, enhancing your overall experience.

To maximize efficiency, familiarize yourself with the shortcuts and quick access features built into ChatGPT. Keyboard shortcuts are a fantastic way to speed up your workflow, reducing the time spent navigating the interface. For example, pressing 'Ctrl + N' can quickly start a new conversation, while 'Ctrl + S' saves your current chat. These shortcuts are designed to keep your hands on the keyboard, minimizing distractions from your main task. Quick access features, like the ability to pin frequently used tools or conversations, further streamline your process. By pinning important chats, you can jump back into them without searching through menus. These efficiency tips ensure that you spend less time managing the interface and more time focusing on the content of your interactions. As you become more familiar with these tools, you'll find that they significantly enhance your productivity, allowing you to work smarter, not harder.

## 2.2 OVERCOMING TECH ANXIETY: A BEGINNER'S PATH

For many, the thought of interacting with new technology, like ChatGPT, can induce a fair amount of anxiety. This unease often stems from a fear of breaking the system, a common concern for those unaccustomed to digital tools.

You might worry that a wrong click or accidental command could disrupt the AI or cause irreversible damage. Rest assured, these tools are designed with user error in mind. They are robust and forgiving, allowing for experimentation without the risk of catastrophic failure. Additionally, concerns about data security often loom large. It's natural to feel wary about how your information is used or stored. Fortunately, ChatGPT, like many reputable AI services, has stringent security measures in place to protect your data, offering peace of mind for those hesitant to dive in.

Embracing a growth mindset is a powerful antidote to tech anxiety. This mindset encourages you to view challenges as opportunities for learning rather than insurmountable obstacles. With technology, mistakes are part of the process, providing valuable lessons that drive improvement. Perhaps you misinterpret a feature or input a command incorrectly. Instead of viewing these as failures, recognize them as steps toward mastery. Each small victory, such as successfully completing a task or solving a minor issue, is a cause for celebration. These victories build confidence and reinforce the belief that you can navigate new technology effectively. The journey of learning may have its bumps, but each experience contributes to your growing competence and comfort with the tool.

Support resources are vital in easing the transition into using ChatGPT. A wealth of online forums exists where beginners and experts alike exchange tips and solutions. These communities are invaluable, offering a platform to ask questions and share experiences. Engaging with others who are on a similar path can provide reassurance and practical insights. Additionally, tutorial videos and guides are readily available, offering step-by-step instructions for various tasks. These resources serve as a visual aid, making complex processes more accessible. Whether you prefer written guides or video demonstrations, there's a resource out there to match your learning style. These materials can demystify many aspects of ChatGPT, turning what once seemed daunting into manageable and even exciting endeavors.

Practical experience is one of the best ways to build confidence with new technology. By engaging in step-by-step practice scenarios, you can familiarize yourself with ChatGPT's capabilities in a controlled, low-pressure environment. Start with simple tasks that require minimal input, allowing you to focus on understanding the system's responses and features. As you grow more comfortable, introduce interactive exercises that challenge you to apply what you've learned in novel ways. These exercises not only reinforce your knowledge but also encourage exploration and experimentation. By gradually increasing the complexity of tasks, you build a solid foundation of skills, paving the way for more advanced applications. This hands-on approach is crucial in transforming apprehension into confidence, empowering you to use ChatGPT with ease and creativity.

## 2.3 BUILDING CONFIDENCE WITH CHATGPT BASICS

Starting with simple tasks is a great way to build initial confidence when using ChatGPT. Imagine you're about to draft a basic email. Begin by typing out your message in clear, simple sentences. ChatGPT can help refine your wording, suggest formal or casual tones, and even check for grammatical errors. It's like having a personal editor at your fingertips, guiding you to craft the perfect message. Another straightforward task is creating a shopping list. You can jot down the items you need, and ChatGPT will organize them or even suggest categories and quantities based on your input. These tasks might seem mundane, but they serve as a gentle introduction to the capabilities of AI, allowing you to become familiar with its responses and suggestions without feeling overwhelmed.

As you grow more comfortable with these basics, it's time to explore more complex uses.

Consider writing a short blog post. Start with a general topic, and let ChatGPT assist in structuring your thoughts. It can help generate an outline, suggest catchy headlines, and even offer content ideas to make your post engaging. Another intriguing task is generating creative story ideas. Provide ChatGPT with a seed—a character, a setting, or a plot twist—and watch as it spins a narrative web around your initial idea. These exercises not only showcase ChatGPT's versatility but also encourage you to stretch your creative muscles. Over time, as you practice these more advanced tasks, you'll notice an increase in confidence and proficiency, paving the way for even more ambitious projects.

Encouraging experimentation is key to unlocking the full potential of ChatGPT. Don't hesitate to try different prompt styles. Vary your questions, add context, or specify the tone you're aiming for. ChatGPT thrives on variety, and experimenting with different inputs can yield surprising and delightful results. Explore diverse applications, from drafting business proposals to crafting poetry. Each new experiment provides insight into how ChatGPT interprets and responds to different challenges. This freedom to explore fosters a sense of discovery and innovation, empowering you to use ChatGPT creatively and effectively. By embracing experimentation, you not only learn more about the tool but also about your preferences and style, making each interaction more tailored and meaningful.

Feedback and adjustment are integral to refining your interactions with ChatGPT. As you engage with the AI, take note of its responses. Are they clear and relevant? If not, consider how you might adjust your prompts for better clarity. ChatGPT learns from the feedback loop created by your inputs and its responses. By interpreting these interactions, you can fine-tune your prompts to achieve more precise outcomes. Adjusting prompts might involve providing more context, simplifying your language, or specifying the type of response you're seeking. This iterative process enhances your understanding of how ChatGPT operates and improves the quality of your interactions, making the AI a more effective tool in your arsenal.

As you become more adept at using ChatGPT, you'll find that these foundational skills open up a world of possibilities. What began as simple tasks now serve as the stepping stones to more complex applications, from crafting long-form content to engaging in sophisticated problem-solving. This chapter emphasizes the importance of starting small, experimenting freely, and continually refining your approach. By doing so, you set the stage for a productive and innovative partnership with AI. As we conclude this exploration of ChatGPT basics, remember that each interaction builds your confidence and competence. Together, these elements create a robust framework for future learning and application.

# CHAPTER THREE

## CRAFTING EFFECTIVE PROMPTS

Picture this: you're sitting at your desk, faced with a daunting task that requires precision and creativity. You decide to enlist the help of ChatGPT, eager to see how this AI can assist. As you type your first prompt, you're met with a response that feels off the mark. This moment, familiar to many beginners, highlights the art of crafting effective prompts—a skill that turns AI into a powerful ally rather than an enigmatic tool. This chapter delves into the nuances of speaking AI's language, transforming your interactions into meaningful dialogues.

Understanding AI language expectations is crucial for maximizing the potential of ChatGPT. When engaging with AI, clarity and specificity emerge as paramount. Think of the AI as a diligent assistant who thrives on clear instructions. When you use concise language, you streamline the communication process, reducing ambiguity and enhancing the AI's ability to understand and respond. Avoiding vague wording is equally important. Just as a map guides a traveler, explicit prompts guide the AI, ensuring it stays on course without veering into unintended territories. By providing precise directives, you set the stage for a productive exchange, minimizing the risk of misunderstandings and enhancing the quality of the AI's output.

The structure of your prompts plays a significant role in how effectively ChatGPT interprets your requests. Consider breaking down complex instructions into manageable components, akin to assembling a puzzle piece by piece. This approach not only aids the AI in comprehending your intent but also ensures that no crucial detail is overlooked. Prioritizing key information within your prompt is another strategy that enhances clarity. By placing the most critical elements at the forefront, you guide the AI's focus, directing its attention to what truly matters. This structured approach mirrors the process of outlining a plan before executing it, providing a clear framework for the AI to follow.

Various prompt formats exist to facilitate effective communication with AI, each serving a distinct purpose. Question-based prompts, for example, encourage the AI to provide specific answers, making them ideal for gathering information or seeking clarification. Conversely, instructional prompts offer a different dynamic, guiding the AI to perform tasks or generate content based on your directives. By experimenting with these formats, you gain insight into how ChatGPT responds under different conditions, allowing you to tailor your approach to suit your needs. This versatility in prompt crafting empowers you to adapt your communication style, ensuring that the AI aligns with your objectives.

The process of iterative prompt refinement is a cornerstone of effective interaction with ChatGPT. As you engage with the AI, view each exchange as an opportunity to learn and improve. Analyzing AI feedback provides valuable insights into how your prompts are being interpreted. This feedback loop allows you to adjust prompt wording, fine-tuning your language for optimal results. Consider this process akin to sculpting a piece of art, where each refinement brings you closer to the desired outcome. Over time, this iterative approach cultivates a deeper understanding of AI dynamics, enabling you to craft prompts that consistently yield high-quality responses.

## 3.1 REFLECTIVE EXERCISE: PROMPT CRAFTING PRACTICE

Reflect on a recent interaction with ChatGPT. Identify the prompt you used and consider how it could be refined for clarity and specificity. Rewrite the prompt using the principles discussed in this chapter, focusing on breaking down complex instructions, prioritizing key information, and choosing an appropriate format. Compare the AI's response to your original prompt with the revised version, noting any differences in quality and relevance. This exercise encourages continuous improvement and reinforces the techniques covered, allowing you to apply them in real-world scenarios.

By mastering the language of prompts, you unlock the true potential of ChatGPT, transforming it from a simple tool into a dynamic collaborator. As you progress, these skills will serve as the foundation for more advanced interactions, opening doors to new possibilities and insights.

## 3.2 DESIGNING PROMPTS FOR CREATIVE OUTPUTS

Imagine sitting down to write a story, your mind a blank canvas waiting to be filled with vibrant colors and intricate details. Crafting prompts for ChatGPT can be much the same—a delicate blend of structure and imagination that encourages creativity. To stimulate ChatGPT's creative potential, consider the power of open-ended questions. These questions act like springboards, launching the AI into a realm of possibilities where the only limits are your imagination. Instead of asking, "What should I write about?" try posing, "What would happen if a city suddenly appeared overnight?" This kind of prompt invites the AI to explore new ideas, drawing from its extensive knowledge base to offer unique suggestions.

Incorporating imaginative scenarios is another technique to spark creativity. Picture a world where trees whisper secrets to those who listen or a universe where time flows backward. By weaving these elements into your prompts, you provide ChatGPT with a rich tapestry to work from, encouraging it to generate responses that are as vivid and unexpected as they are insightful. These scenarios do not bind the AI to a single path but rather open multiple avenues for exploration. The beauty of such prompts lies in their ability to transcend conventional thinking, offering fresh perspectives and novel solutions to familiar problems.

Balancing structure and freedom within prompts is crucial for guiding ChatGPT's creative output without stifling its potential. Think of this balance as providing a framework while allowing flexibility. You might set the stage with a specific setting or theme but leave the details open for interpretation. For example, you could prompt, "In a world where technology controls nature, describe a day in the life of a farmer." Here, the structure is evident, yet there is ample room for creative exploration. Setting boundaries for creative exploration ensures the AI remains focused on the task while still having the freedom to innovate.

Examples of creative prompts highlight how this balance can yield exciting results. Storytelling prompts like "Describe a journey through a land where emotions are tangible objects" offer a narrative framework while encouraging ChatGPT to delve into rich details. Similarly, brainstorming session prompts might ask, "What inventions could help humanity thrive in a post-apocalyptic world?" These prompts push the AI to think beyond the ordinary, generating ideas that may inspire your own creative ventures. By experimenting with different prompts, you can discover which techniques best stimulate the AI's creative faculties.

Evaluating the creativity and originality of AI-generated content is a nuanced task that requires a keen eye. When assessing these outputs, consider criteria such as novelty, coherence, and relevance to the prompt. A creative response should offer new insights or perspectives, presenting ideas that are not immediately obvious. Coherence is equally important, ensuring that the response flows logically and maintains consistency throughout. Relevance ties back to the prompt, confirming that the AI has addressed the core question or scenario presented. Establishing feedback loops for creativity enhancement is a valuable practice. By providing feedback on what aspects of the response were particularly effective or where improvements could be made, you guide ChatGPT towards producing even more refined outputs in future interactions.

The iterative process of crafting and evaluating creative prompts is akin to nurturing a garden. Each prompt is a seed planted with care, tended through feedback and refinement, and eventually blossoming into a unique creation. As you engage with ChatGPT, you'll find that this dynamic interaction fosters a deeper understanding of how to elicit the AI's creative potential. By blending open-ended exploration with structured guidance, your prompts become not just questions or commands, but collaborative efforts that unlock the full spectrum of ChatGPT's capabilities, transforming your creative process into a vibrant and ever-evolving narrative.

## 3.3 TROUBLESHOOTING COMMON PROMPT ERRORS

When working with ChatGPT, certain prompt errors can lead to frustrating and unexpected results. One common mistake is using vague language, which can cause the AI to produce unclear or irrelevant responses. Imagine asking someone for advice without specifying the context—you're likely to receive a broad answer that misses the mark. Similarly, ChatGPT thrives on specificity. Without it, the AI struggles to pinpoint exactly what you're seeking, resulting in outputs that may not meet your expectations. Another frequent error involves overloading prompts with too much information. While it might seem like providing more details would help, it often confuses the AI, making it difficult to focus on the core request. Just as a cluttered room can hinder productivity, an overloaded prompt can overwhelm ChatGPT, leading to scattered and less coherent replies.

To correct these errors, start by simplifying complex prompts. Break down your request into smaller, more manageable components, each focusing on a single aspect. This method helps clarify your intentions and makes it easier for ChatGPT to follow along. Think of it as offering a clear path rather than a tangled web. Additionally, keep your prompts focused on key objectives. Identify the primary goal of your interaction and ensure that your prompt aligns with that purpose. This focus prevents the AI from wandering off-course and helps maintain the relevance of its responses. By honing in on what truly matters, you guide ChatGPT to provide more accurate and useful outputs.

AI misinterpretations, while sometimes frustrating, offer valuable learning opportunities. When ChatGPT delivers an unexpected or incorrect response, examine it closely. Analyze what might have led to the misinterpretation: Was the prompt too broad? Did it lack necessary context? By identifying the root of the misunderstanding, you can adjust your prompts to be more effective in the future. This process of refinement is akin to iterating on a design until it functions optimally. Adjusting prompts based on AI responses not only improves the quality of interactions but also enhances your understanding of how ChatGPT processes information. Over time, this iterative approach leads to more fruitful exchanges, as you learn to communicate more effectively with the AI.

### Case Study: Prompt Optimization in Action

Consider a scenario where a small business owner uses ChatGPT to draft marketing emails. Initially, the prompts are vague, asking the AI to "write a promotional email for a new product." The results are generic and lack the punch needed to engage customers. Recognizing the need for improvement, the owner refines the prompt to include specific details about the product's unique features and target audience. With this focused prompt, ChatGPT generates a more compelling and tailored email that resonates with recipients. This real-world example illustrates how prompt optimization transforms AI interactions, turning ordinary exchanges into impactful and efficient collaborations.

As you become adept at recognizing and addressing these common prompt errors, you'll find that your interactions with ChatGPT become more productive and satisfying. By embracing the learning opportunities that arise from AI misinterpretations and applying strategies to refine your prompts, you pave the way for more effective communication. Each adjustment brings you closer to mastering the nuances of AI language, enhancing the overall quality of your exchanges. As we explore these techniques, remember that prompt optimization is an ongoing process, one that continually evolves as you gain experience and insight.

## 3.4 ADVANCED PROMPT TECHNIQUES FOR BETTER INTERACTION

Imagine you're in a café, trying to explain your next big project to a friend. You wouldn't just blurt out random ideas without context, right? You'd start by setting the scene, maybe sharing a backstory, and tying it all together so your friend can follow along. Similarly, when crafting prompts for ChatGPT, incorporating contextual information can significantly enhance the effectiveness of your interactions. By embedding background details into your prompts, you provide the AI with a richer tapestry of information to draw from, leading to more relevant and nuanced responses. For instance, if you're asking ChatGPT to draft a marketing email, mentioning the target audience's preferences or a recent campaign's outcomes can guide the AI to craft a message that resonates well. Additionally, linking prompts for continuity can maintain the flow of conversation, ensuring that ChatGPT doesn't treat each interaction in isolation but rather as a part of an ongoing dialogue.

Handling complex tasks with AI often requires dynamic prompting techniques. Imagine you're guiding a group through a multi-step process; you wouldn't lay out every detail all at once. Instead, you'd introduce each step sequentially, allowing time for questions and clarifications. In the same vein, layered prompting methods can break down intricate tasks into manageable chunks, enabling ChatGPT to address each component thoroughly before moving to the next. This approach not only fosters clarity but also allows you to use follow-up questions to refine the AI's responses, ensuring that each answer is aligned with your overall objective. For example, if you're using ChatGPT to analyze a detailed report, you might start by asking it to summarize key findings, then follow up with questions about specific data points or trends. This method encourages a deeper, more comprehensive exploration of the task at hand. Engagement is key in any interaction, whether with a person or an AI. To keep users engaged, interactive prompt strategies can transform a stagnant exchange into a dynamic dialogue. Consider using interactive dialogue prompts that invite ChatGPT to respond in a conversational manner, encouraging back-and-forth exchanges that mimic a real conversation. This technique not only holds the user's attention but also enriches the interaction, making it more informative and enjoyable. Moreover, prompt-response feedback loops can enhance AI interaction by creating a cycle of continuous improvement. By regularly evaluating the AI's outputs and adjusting prompts accordingly, you foster an environment where both the user and the AI learn from each exchange, leading to increasingly refined responses over time.

To support the development of advanced prompt techniques, a variety of tools and resources are available. Prompt optimization tools, for instance, can assist in crafting more effective prompts by providing insights into language patterns and response tendencies. These tools help identify areas for improvement, offering suggestions to enhance clarity and relevance. Additionally, community resources for prompt engineering offer a wealth of shared knowledge and expertise. Engaging with online forums and discussion groups can provide valuable tips and tricks from others who have mastered the art of prompt crafting. These resources foster a collaborative environment where you can learn from the experiences of others, refining your skills and expanding your understanding of AI interaction.

As we wrap up this chapter on crafting effective prompts, it's clear that mastery in this area opens new avenues for leveraging ChatGPT's potential. By incorporating context, employing dynamic prompting techniques, and utilizing interactive strategies, you transform your interactions with AI into powerful tools for innovation and problem-solving. The advanced techniques discussed here provide a foundation for deeper and more meaningful engagement with ChatGPT, setting the stage for exploring its applications in everyday tasks and beyond.

# CHAPTER FOUR

## PRACTICAL APPLICATIONS IN DAILY TASKS

Picture a morning where, instead of being greeted by a mountain of unread emails, you find your inbox neatly organized, with messages sorted and replies ready to send. This is not a fantasy but a practical application of ChatGPT, an AI tool that can revolutionize your email management. Emails, though essential for communication, can become overwhelming, consuming valuable time and energy. In fact, the average worker spent nearly 149 minutes per day managing emails in 2021. With ChatGPT, you can transform how you handle emails, making the process faster, more efficient, and less stressful.

One of the most time-consuming aspects of email management is responding to routine inquiries. Whether it's answering frequently asked questions or acknowledging orders, these tasks can eat into your day. ChatGPT can automate these responses, saving you precious time. By creating template responses for FAQs, you ensure that every customer receives timely and consistent information. Additionally, setting up auto-replies for common inquiries allows you to maintain communication even when you're away from your desk. This automation doesn't just keep your inbox in check; it also enhances customer satisfaction by providing immediate responses, ensuring that no message goes unanswered.

Drafting professional emails often requires careful consideration of tone and structure. ChatGPT excels in this area, enabling you to craft well-structured messages quickly. By outlining your email content using bullet points, you organize your thoughts clearly and concisely. This method helps you focus on key points, ensuring that your message is both informative and direct. Moreover, ChatGPT can adjust the tone and formality to match recipient expectations, whether you're addressing a colleague or reaching out to a potential client. This adaptability ensures that your communication remains professional and appropriate, regardless of the context.

In addition to drafting, personalizing email correspondence is crucial for building relationships and conveying sincerity. ChatGPT can assist in incorporating recipient-specific details, making each email feel tailored without losing efficiency. For instance, you might include the recipient's recent project achievements or personal interests, demonstrating your attention to detail and fostering a sense of connection. Tailoring content to match the recipient's interests not only personalizes the interaction but also increases the likelihood of engagement. This balance between personalization and efficiency is key to maintaining effective communication in today's fast-paced world.

Managing email overload is another challenge that many face. With the constant influx of messages, it can be difficult to prioritize and respond promptly. ChatGPT offers strategies to manage and prioritize your inbox effectively. By categorizing emails based on urgency, you can focus on what truly matters, addressing high-priority messages first. Additionally, ChatGPT's ability to filter promotional content automatically ensures that your primary communication remains front and center, reducing distractions and clutter. This streamlined approach not only saves time but also reduces stress, allowing you to tackle your inbox with confidence and clarity.

**Interactive Element: Email Management Checklist**

- Set up template responses for frequently asked questions.

- Draft auto-replies for common inquiries to maintain communication.
- Use bullet points to outline professional email content.
- Adjust tone and formality to match recipient expectations.
- Incorporate recipient-specific details for personalized emails.
- Categorize emails based on urgency to focus on priorities.
- Filter promotional content automatically to reduce inbox clutter.

By integrating these strategies, you harness ChatGPT's capabilities to streamline your email communication, transforming it from a burden into an efficient and manageable task. This chapter delves into the practical ways ChatGPT can enhance your daily work, offering solutions that are both accessible and effective.

## 4.1 BRAINSTORMING CREATIVE IDEAS WITH AI

In the realm of creativity, fresh and innovative ideas often emerge from the interplay of imagination and diverse thinking techniques. ChatGPT serves as a catalyst in this creative process, opening doors to new possibilities. When you engage with ChatGPT using divergent thinking techniques, the

AI can help explore avenues you might not have considered. Consider prompting it with "what if" scenarios—these questions act as springboards into the unknown, challenging the AI to generate ideas that defy conventional limits. Imagine asking, "What if buildings could grow like trees?" Such prompts encourage ChatGPT to dive into realms of creativity, offering suggestions that can spark your own imagination and lead to groundbreaking concepts. By engaging in this type of dialogue, you're not just generating ideas but fostering a mindset that embraces innovation, pushing the boundaries of what seems possible.

Collaboration is another pillar of creativity, and ChatGPT excels in facilitating collaborative brainstorming sessions. By co-creating mind maps with AI input, you and your team can visualize ideas in an organized and dynamic format. This approach allows everyone to see the connections between different concepts, encouraging the flow of ideas and enhancing group creativity. ChatGPT can also incorporate diverse perspectives through AI suggestions, offering insights that may not be immediately apparent to the human mind. For instance, when brainstorming for a new product, ChatGPT might suggest features or uses based on global trends or historical data that add depth to the discussion. This diversity of thought can lead to more comprehensive solutions and a richer creative process, transforming a simple session into a dynamic exploration of possibilities.

Creative blocks are a common hurdle, but ChatGPT can help you overcome them by offering fresh angles and perspectives. Random prompts from ChatGPT can trigger inspiration by presenting ideas or questions that disrupt stagnant thinking patterns. Imagine you're stuck on a project, and ChatGPT suggests, "What if you approached this from a child's perspective?" This shift in viewpoint can unlock new paths and ideas, reigniting your creative spark. Additionally, ChatGPT can explore unconventional solutions that challenge the status quo, encouraging you to think outside the box. By embracing these unexpected suggestions, you might discover innovative approaches to problems, turning obstacles into stepping stones for creativity.

Once you have a pool of ideas, refining and expanding them is crucial for developing them into feasible projects. ChatGPT can assist in developing detailed outlines from basic concepts, helping you organize thoughts into coherent plans. With a structured framework, you can assess the viability of each idea, making necessary adjustments to strengthen its potential. Furthermore, ChatGPT can explore potential challenges and solutions, providing a comprehensive view of the project's landscape. This foresight allows you to anticipate obstacles and devise strategies to overcome them, increasing the likelihood of success. By collaborating with ChatGPT, you refine your ideas and enhance their depth, ensuring that each concept is thoroughly vetted and ready for implementation.

**Reflective Exercise: Idea Expansion**

Choose one idea you've generated with ChatGPT. Break it down into its core components and use ChatGPT to explore each aspect in detail. Ask questions about potential challenges, target audiences, and implementation strategies. Note the AI's responses and consider how they refine your understanding of the idea's strengths and weaknesses. This exercise encourages you to critically evaluate and develop your concept, transforming an initial spark into a well-rounded plan.

## 4.2 QUICK RESEARCH HACKS USING CHATGPT

Imagine you're tasked with a research project, and the clock is ticking. Gathering information quickly becomes a priority, and this is where ChatGPT shines. It can swiftly summarize lengthy articles, pulling out the essence of texts that might otherwise take hours to dissect. Whether you're skimming through academic papers or news articles, ChatGPT identifies key data points, transforming dense material into concise summaries. This ability to distill information not only saves time but also helps you focus on the most relevant facts, making the research process less daunting and more efficient. You no longer need to wade through pages of information; instead, you have a tool that does the heavy lifting, allowing you to concentrate on analysis and application.

Cross-referencing multiple sources is crucial for ensuring the accuracy and reliability of your research. ChatGPT aids this process by comparing different study results, highlighting both consistencies and discrepancies. It helps validate facts against trusted databases, acting as a second pair of eyes that ensures your data's integrity. This cross-referencing capability is particularly valuable in fields requiring precision, such as scientific research or historical analysis. By leveraging ChatGPT, you can confidently build a foundation of reliable information, making your conclusions more robust and your arguments more compelling. The AI's ability to scan vast datasets and synthesize them into coherent summaries streamlines the validation process, reinforcing the credibility of your work.

When faced with complex information, synthesizing it into understandable summaries is often a challenge. ChatGPT excels in this domain by creating easy-to-digest summaries that make sense of convoluted data. It translates technical jargon into layman's terms, making specialized knowledge accessible to everyone. Whether you're navigating through scientific reports or financial analyses, ChatGPT ensures that the core message is clear and comprehensible. This translation of complexity into simplicity not only aids understanding but also enhances retention, empowering you to communicate findings effectively to diverse audiences. The ability to demystify complex topics is invaluable in education, presentations, and collaborative projects, where clarity is paramount.

Organizing research findings is another area where ChatGPT proves invaluable. It helps create annotated bibliographies, categorizing sources with notes on their relevance and contribution to your research. This organization makes it easier to track your sources and reference them accurately, ensuring a thorough and well-documented project. Additionally, ChatGPT assists in developing thematic folders for research topics, grouping related information for easy retrieval. This systematic approach to organization simplifies the research process, allowing you to focus on synthesis and analysis rather than searching for scattered information. By structuring your data efficiently, ChatGPT enhances the coherence and depth of your research, making your final output both informative and persuasive.

**Textual Element: Resource List**

- Use ChatGPT to summarize articles from trusted sources like Google Scholar or JSTOR.
- Validate data against databases such as PubMed for scientific research.
- Translate industry jargon into simple language with ChatGPT for presentations.
- Create annotated bibliographies with tools like Mendeley or EndNote, using ChatGPT for annotations.
- Develop thematic research folders on platforms like Notion or Evernote, with ChatGPT aiding in content categorization.

## 4.3 ENHANCING PRODUCTIVITY WITH AUTOMATED TASKS

In a world where every second counts, automating repetitive tasks can dramatically increase your productivity. ChatGPT is a versatile tool that helps streamline these everyday tasks, freeing up your time for more strategic activities. Imagine waking up and finding your day already organized, with reminders set and tasks arranged. ChatGPT can handle scheduling with ease, setting reminders for meetings, deadlines, and even personal events. This AI takes the mental load off your shoulders, ensuring you never miss an important appointment or task. Additionally, automating data entry processes becomes a breeze with ChatGPT's capabilities. It can extract information from various sources and input it into your databases, saving you countless hours of tedious work. This automation not only increases efficiency but also reduces errors, as the AI meticulously follows instructions without deviation.

Integrating ChatGPT into your daily workflow can optimize efficiency, allowing you to focus on high-priority tasks. Consider how much time you spend managing projects—coordinating with team members, tracking progress, and updating tasks. ChatGPT streamlines project management by organizing tasks, setting deadlines, and even sending reminders to team members. This organization ensures that everyone stays on track and meets deadlines. Furthermore, ChatGPT can reduce time spent on administrative tasks, such as updating records or preparing reports. By automating these processes, you can redirect your efforts toward strategic thinking and innovation. The AI's ability to manage multiple tasks simultaneously enhances your workflow, making it more fluid and less fragmented.

Tracking and analyzing productivity are key to understanding how efficiently you're working. ChatGPT can generate detailed productivity reports, offering insights into how you allocate your time. These reports highlight where you're most productive and where improvements can be made. For instance, you might discover that you spend too much time on low-priority tasks. With this data, you can adjust your schedule to focus on activities that offer the highest return. Additionally, ChatGPT can analyze time spent on various tasks, helping you identify patterns and bottlenecks. By understanding these elements, you can implement strategies to improve your workflow, such as delegating tasks or setting stricter time limits. This analysis fosters a culture of continuous improvement, where each day becomes an opportunity to refine and enhance your approach.

Continuous improvement is vital for maintaining productivity in the long term. ChatGPT aids this by helping you set achievable productivity goals. By outlining specific objectives, you create a roadmap for success, giving you clear targets to aim for. These goals provide direction and motivation, ensuring that you remain focused and driven. Moreover, ChatGPT facilitates feedback loops for self-assessment. After completing tasks, you can review your performance and make adjustments as needed. This self-reflection is crucial for identifying strengths and weaknesses, allowing you to capitalize on the former and address the latter. By incorporating these feedback loops, you develop a proactive approach to productivity, consistently seeking ways to enhance your performance. This ongoing process of evaluation and adaptation ensures that you remain agile and responsive to changing demands, optimizing your productivity over time.

**Textual Element: Productivity Improvement Checklist**

- ❖ Automate scheduling and data entry tasks with ChatGPT.
- ❖ Use ChatGPT to organize project management tasks and set reminders.
- ❖ Generate productivity reports to analyze time allocation.

- ❖ Set specific productivity goals with ChatGPT's assistance.
- ❖ Implement feedback loops for self-assessment and continuous improvement.

As we wrap up this chapter, consider how integrating ChatGPT into your daily tasks can not only enhance productivity but also open doors to new possibilities. Whether it's automating repetitive tasks, optimizing workflows, or tracking productivity, ChatGPT offers tools that transform how you work. In the next chapter, we'll explore how AI can assist in overcoming challenges and maximizing effectiveness.

# CHAPTER FIVE

## OVERCOMING COMMON CHALLENGES

Imagine walking into a library where books are piled high, spilling from every corner, with new volumes arriving by the minute. You stand in awe, overwhelmed by the sheer volume, unsure where to begin. This flood of information mirrors the experience many face when interacting with AI tools like ChatGPT. Information overload can feel paralyzing, transforming a helpful resource into a source of stress. With AI's capability to generate vast amounts of content, managing this influx becomes crucial. You may find yourself bombarded with frequent updates, notifications, and an ever-growing list of AI-generated possibilities. Recognizing the sources of this overload is the first step in taming it.

One common culprit is the constant stream of AI-generated updates. Every time you interact with ChatGPT or similar tools, new data is produced, creating a cycle that can quickly become overwhelming if not managed properly. This continuous flow of information, while valuable, can lead to decision fatigue, where the sheer volume makes it difficult to focus on what's important. Additionally, the overwhelming amount of generated content can clutter your workspace, making it challenging to identify actionable insights. These challenges, often compounded by the absence of a solid information management foundation, can hinder your ability to extract meaningful insights from the data at hand.

Streamlining your information intake can significantly alleviate the burden of overload. One effective technique is setting filters and priorities. By identifying the most relevant data and prioritizing it, you can focus on what truly matters without getting lost in the noise. This approach mirrors the act of sorting mail, where you separate essential documents from promotional flyers. Utilizing summary features can also condense information, allowing you to grasp the core message without wading through pages of content. This distillation is akin to extracting the essence of a novel in a single paragraph, providing clarity and reducing cognitive strain. These techniques, when implemented thoughtfully, shift your focus from quantity to quality, enhancing your engagement with AI-generated outputs.

Organizing digital content efficiently is another vital strategy. Consider creating categorized folders for AI outputs, much like organizing files in a cabinet. This method ensures easy access and retrieval, transforming chaos into order. Utilizing tagging systems can further streamline your digital workspace, allowing you to label and sort content based on themes or relevance. Tags act as virtual bookmarks, guiding you to the information you need without sifting through unrelated material. This organization not only saves time but also enhances productivity, enabling you to navigate the digital landscape with ease. By establishing a structured system for managing AI-generated content, you create a foundation for efficient information management.

Limiting unnecessary interactions with AI is crucial to avoid cognitive overload. Setting specific interaction goals can help maintain focus, ensuring that each engagement with ChatGPT serves a clear purpose. This approach prevents aimless exploration and encourages deliberate use, much like setting an itinerary for a trip. Additionally, scheduling designated times for AI consultations can provide structure, balancing interaction with periods of rest and reflection. This practice mirrors the concept of time blocking, where specific tasks are allocated distinct time slots, optimizing productivity. By consciously managing the frequency and purpose of your AI interactions, you create a sustainable approach that maximizes benefits while minimizing stress.

### Reflective Exercise: Information Management Plan

Take a moment to assess your current interaction with AI tools. Identify three sources of information overload you encounter regularly. Develop a plan to manage these sources, incorporating techniques such as setting filters, creating organized folders, and scheduling designated interaction times. Reflect on how these changes might impact your experience with AI, noting any anticipated challenges or benefits. This exercise encourages proactive information management, empowering you to tailor your AI engagement to your needs.

## 5.1 FINDING RELIABLE SOURCES WITH AI ASSISTANCE

Imagine you're tasked with writing a report. You turn to ChatGPT for help, but how can you be sure the information it provides is credible? Evaluating the credibility of AI-suggested sources is a crucial skill. Start by cross-checking AI-generated claims with established databases. This step is like having a second opinion, ensuring the information is grounded in reality. Assessing author credentials is equally important. Consider their expertise and the reputation of the publication. Reliable sources typically have authors with proven backgrounds and are published in reputable outlets. This process, much like vetting a job applicant, ensures you're working with trustworthy data.

Leveraging ChatGPT to recommend reliable sources can be a game-changer. AI algorithms excel at source vetting, filtering through vast amounts of information to suggest credible options. You can customize these recommendations based on your preferences, tailoring them to your specific needs. This customization transforms AI from a generalist into a personalized assistant, honing in on areas that matter most to you. For instance, if you're researching climate science, instruct ChatGPT to prioritize peer-reviewed journals or government reports, ensuring the information aligns with rigorous standards. This strategic use of AI recommendations streamlines your research process, enhancing both efficiency and accuracy.

Integrating verification tools with ChatGPT further bolsters the reliability of your research. Fact-checking tools and plug-ins can be linked to your AI interactions, providing real-time verification of the information you receive. These tools act as digital editors, scrutinizing data for inconsistencies or errors. Multi-source comparisons enhance this process, cross-referencing AI outputs with multiple credible sources to validate claims. This practice is akin to triangulating a position on a map, using multiple points to ensure you're on solid ground. By incorporating these verification methods, you create a robust framework for evaluating AI-generated content, reducing the risk of relying on inaccurate or biased information.

Developing a skeptical mindset towards AI-suggested information is essential for maintaining critical thinking. Question the validity of AI recommendations, just as you would any other source. Engage in independent research alongside AI suggestions, using your judgment to assess the information's reliability. This skepticism is not about distrust but about fostering a healthy critical perspective. It encourages you to consider alternative viewpoints and verify facts before accepting them as truth. By adopting this mindset, you become an active participant in the research process, ensuring that AI serves as a tool for exploration rather than a sole authority.

**Interactive Element: Source Evaluation Checklist**

- ❖ Cross-check AI-generated claims with established databases for accuracy.

- ❖ Assess author credentials and publication reputation to ensure credibility.
- ❖ Customize AI source recommendations based on specific preferences.
- ❖ Integrate fact-checking tools and use multi-source comparisons for verification.
- ❖ Maintain a skeptical mindset, questioning AI suggestions and conducting independent research.

Through these practices, you can confidently navigate the vast landscape of AI-generated information, transforming ChatGPT from a simple tool into a powerful ally in your quest for knowledge.

## 5.2 AVOIDING COMMON PITFALLS IN AI INTERACTION

When you first begin using ChatGPT, you may be amazed at its ability to generate responses and assist with tasks. However, it's important to be aware of common pitfalls that can arise in these interactions. One such challenge is the misinterpretation of nuanced prompts. You might ask a question or give a command that seems straightforward to you, but the AI, lacking human intuition, may misinterpret your intent. This misunderstanding often occurs when prompts lack clarity or include subtlety that the AI cannot recognize. For instance, if you ask for a creative take on a familiar topic, the AI might produce something generic if the prompt doesn't specify the desired angle or tone. These misinterpretations remind us that while AI is advanced, it still requires precise input to function as expected.

Another potential pitfall is an overreliance on AI for decision-making. It's easy to lean heavily on ChatGPT, especially when it can handle tasks quickly and efficiently. However, it's crucial to remember that AI should complement, not replace, human judgment. Dependence on AI might lead you to bypass critical thinking, which can be risky if the AI's response lacks context or depth. For example, using ChatGPT to draft an important email without reviewing it may result in a message that doesn't fully capture the intended tone or content. This reliance can also stifle creativity, as you might default to AI-generated ideas instead of exploring your own. Recognizing the AI's role as an assistant rather than a decision-maker ensures that you remain an active participant in the process, using your judgment to guide the outcome.

To minimize misunderstandings with ChatGPT, establishing clear communication guidelines is invaluable. Setting clear expectations for AI outcomes can significantly improve the quality of interactions. Before you start, consider what you want from the AI and articulate that in your prompt. Providing context is equally important. By giving the AI background information, you enable it to generate responses that are more aligned with your needs. For instance, if you're asking for help with a project, mention previous steps or specific goals to provide a framework for the AI's response. These guidelines help bridge the gap between human intention and AI interpretation, creating a more productive exchange.

Balancing AI interaction with human input is essential for optimal results. While ChatGPT can offer insights and suggestions, evaluating these before implementation is crucial. This evaluation involves assessing the AI's recommendations against your knowledge and experience. By doing so, you ensure that the final decision reflects both the AI's capabilities and your expertise. Incorporating human creativity and insights into the process also enhances the outcome, as it brings a dimension that AI cannot replicate. For instance, combining AI-generated ideas with your unique perspective can lead to innovative solutions that neither could achieve alone. This balance between AI and human input fosters a dynamic interaction where both contribute to the final product.

Monitoring and adjusting your AI use based on feedback is a practice that can refine your interactions over time. Regularly reviewing AI outputs allows you to assess their effectiveness and relevance. This review helps identify patterns in the AI's responses, highlighting areas where adjustments might be needed. Adapting strategies based on these insights ensures that your use of ChatGPT evolves with your needs. For instance, if you notice the AI frequently misinterprets certain prompts, you can refine your input for better clarity. This ongoing adjustment process enhances the AI's utility, making it a more effective tool for your tasks. By actively engaging in this cycle of feedback and refinement, you maintain control over the interaction, ensuring that it continues to meet your expectations.

## 5.3 HANDLING ETHICAL CONCERNS RESPONSIBLY

When using AI tools like ChatGPT, understanding the ethical implications of your actions is crucial. These implications span a range of issues, from data privacy to the potential for bias in AI applications. In an age where personal information can be easily collected and analyzed, respecting data privacy and securing user consent is more important than ever. You should be mindful of how data is gathered and used, ensuring that all interactions with AI respect individuals' privacy. This respect involves obtaining explicit consent from users before collecting their data and being transparent about how their information will be used. Beyond privacy, there's a pressing need to avoid bias in AI applications. AI systems, including ChatGPT, learn from the data they are trained on. If this data is biased, the AI's outputs can reinforce existing prejudices, leading to unfair or discriminatory outcomes. Therefore, it's imperative to actively seek diverse datasets and challenge biases when they appear, ensuring that AI outputs are fair and representative of varied perspectives.

Implementing ethical practices involves taking tangible steps to ensure that your use of ChatGPT aligns with these principles. Transparency in AI interactions is one such step. You should make it clear when AI is being used, so users are aware they are interacting with a machine rather than a human. This transparency builds trust and ensures that users understand the nature and limitations of the interaction. Prioritizing fair representation in AI-generated content is equally important. AI should be used to amplify diverse voices, not to perpetuate stereotypes or exclude certain groups. By actively monitoring AI outputs and making adjustments to promote inclusivity, you can ensure that the content generated reflects a wide range of experiences and viewpoints. These practices require ongoing vigilance and a commitment to ethical standards, but they are essential for responsible AI use.

Promoting ethical awareness among peers and within communities can further support responsible AI use. Advocacy for responsible AI involves educating others about the potential risks and benefits of AI technologies and encouraging them to consider ethical implications in their work. Participating in ethical AI discussions and forums can be a powerful way to engage with these topics, providing a platform to share knowledge and collaborate on solutions. These forums allow you to stay informed about the latest developments in AI ethics and to contribute to the broader conversation about how AI can be used responsibly. By fostering a culture of ethical awareness, you help ensure that these technologies serve the greater good, benefiting society as a whole rather than causing harm or perpetuating inequality.

Addressing potential ethical dilemmas requires foresight and preparedness. Developing contingency plans for ethical breaches is a proactive step that can mitigate harm if issues arise. These plans should outline clear steps for addressing breaches, including how to communicate with affected parties and rectify any damage. Consulting ethical guidelines and experts can also provide valuable insights into best practices and help you navigate complex situations. By seeking guidance from those with expertise in AI ethics, you can ensure that your actions align with established standards and reflect a commitment to ethical integrity. This proactive approach not only protects individuals and communities but also strengthens the credibility and reliability of AI tools like ChatGPT.

Through understanding, implementing practices, promoting awareness, and addressing dilemmas, you can navigate the ethical landscape of AI responsibly. These steps safeguard against misuse and ensure that AI remains a tool for positive change. As we conclude this chapter, consider how these ethical considerations impact your use of ChatGPT and prepare for the next chapter's exploration of skill development.

# CHAPTER SIX

## DEVELOPING SKILLS WITH CHATGPT

Imagine you're in a lively conversation with a friend where every word counts, and the flow of dialogue feels effortless. Now, picture that friend as ChatGPT, a tool that can be as engaging and insightful, if only you know how to speak its language. In this chapter, we dive into the heart of effective communication with AI: mastering prompt engineering. This skill is the key to unlocking ChatGPT's full potential, transforming your interactions from basic queries into rich, dynamic exchanges. Prompt engineering is about crafting your prompts in such a way that the AI not only understands your request but also provides responses that are clear, accurate, and tailored to your needs. This is where the art meets science, blending creativity with structure to achieve the best results.

Advanced prompt structuring is a method that can significantly enhance AI comprehension and response accuracy. One effective technique involves using conditional prompts for scenario-based responses. This means setting specific conditions within your prompts that guide ChatGPT through various potential outcomes. For example, if you're exploring marketing strategies, you might prompt, "If the target audience increases by 10%, how should we adjust our campaign?" Such conditional prompts allow ChatGPT to consider different variables, offering tailored suggestions based on the given scenarios. Another method is layered prompts for multi-step instructions. Here, you break complex tasks into individual steps, guiding the AI through each phase. If you're planning an event, you might start with, "List the required venues," followed by, "Suggest themes for each venue." This layered approach ensures that ChatGPT processes each part of your request thoroughly, resulting in comprehensive and coherent responses.

Incorporating contextual cues into your prompts can refine their specificity and relevance. This involves embedding background details within prompts to provide ChatGPT with the necessary context for more accurate responses. Imagine you're discussing a novel set in the 1800s; instead of vaguely asking for historical facts, you could prompt, "In the context of 19th-century England, describe the societal impact of the Industrial Revolution." By embedding such details, you give ChatGPT a framework to operate within, enhancing the relevance of its outputs. Sequential prompts also help maintain continuity in conversations. If you're having a multi-part discussion with ChatGPT, you can structure your prompts so that each builds on the previous one, ensuring a cohesive dialogue. This strategy is similar to constructing a narrative, where each chapter logically follows the last, creating a seamless flow of information.

Utilizing feedback for refinement is an iterative process that enhances your prompt crafting skills. By analyzing response patterns, you can adjust prompt phrasing to improve clarity and effectiveness. If you notice that ChatGPT frequently misunderstands your requests, take a closer look at how you're phrasing them. Maybe the language is too complex or the instructions too vague. Adjusting these elements can make a significant difference in the AI's responses. Incorporating user feedback loops is another valuable strategy. After receiving an AI-generated response, reflect on its accuracy and usefulness. Was it what you expected? If not, consider how you might modify your prompt for a better outcome next time. This continuous cycle of feedback and adjustment not only refines your skills but also trains ChatGPT to respond more effectively to your specific needs.

Exploring diverse prompt applications encourages experimentation and creativity in your interactions with ChatGPT. Prompts designed for educational purposes can transform learning experiences by generating study guides, summarizing complex topics, or even posing quiz questions. This approach turns ChatGPT into a versatile educational tool, capable of adapting to various learning styles and needs. Interactive storytelling and role-playing prompts offer another avenue for exploration. You might prompt ChatGPT to create a story where you and the AI co-author the plot, each contributing characters, settings, and twists. This interactive format not only engages your imagination but also showcases ChatGPT's ability to adapt and respond to creative challenges. By experimenting with different prompt types, you expand the horizons of what ChatGPT can do, discovering unique applications that enhance both your personal and professional endeavors.

**Reflective Exercise: Prompt Experimentation**

Take a moment to write down three different types of prompts you might use with ChatGPT. One should be a conditional prompt, another should include contextual cues, and the third should be for an educational purpose. Try these prompts with ChatGPT and observe the responses. Reflect on what worked well and what could be improved. This exercise encourages you to experiment with prompt engineering techniques, enhancing your ability to craft effective and engaging interactions with AI.

## 6.1 ENHANCING CREATIVE WRITING WITH AI

When you sit down to write a story, the world within your mind is vast and full of potential. Yet, translating that world into words can sometimes feel like an uphill battle. This is where ChatGPT steps in, not just as a tool but as a co-author in your creative process. Imagine having an AI that helps you develop character backstories with depth and nuance. You start with a basic outline for a character, perhaps a young detective with a troubled past. ChatGPT can assist by suggesting details that flesh out this character's history, such as childhood influences or pivotal life events that shaped their journey. These suggestions add layers to your narrative, making characters more relatable and compelling. Similarly, when your plot feels stagnant, ChatGPT can help generate plot twists and new story arcs. By feeding it key elements of your story, you can prompt the AI to propose unexpected turns or conflicts, injecting fresh energy into your narrative.

Writer's block is a common adversary, one that can halt creativity in its tracks. Overcoming it requires a shift in perspective, and ChatGPT offers just that. By prompting the AI for alternative narrative perspectives, you can explore the same story through a different character's eyes or from an alternative angle. This not only broadens your storytelling but also reignites your inspiration. For instance, if you're stuck on how to continue a chapter, asking ChatGPT to narrate the scene from another character's viewpoint can open up new possibilities. Additionally, using random prompts to spark inspiration can be incredibly effective. You might input a simple, unrelated prompt like "A stormy night reveals hidden truths," and let ChatGPT weave it into your story. These random elements act as catalysts, breaking the monotony and sparking new ideas that propel your writing forward.

Refining your writing style and tone is another area where ChatGPT proves invaluable. Perhaps you admire the lyrical prose of a famous author and wish to emulate that style. ChatGPT can mimic these styles, allowing you to experiment with different voices in your writing. By analyzing the structure and rhythm of well-known authors, the AI can suggest ways to incorporate similar techniques into your work. This experimentation helps you find your unique voice while learning from the masters. Adjusting tone to match genre conventions is equally important. Whether you're crafting a suspenseful thriller or a whimsical fairy tale, ChatGPT can guide you in maintaining the appropriate tone throughout your narrative. By providing feedback on your existing text, the AI ensures consistency in style, enhancing the overall quality of your work.

Editing and proofreading are crucial final steps in the writing process, and ChatGPT can assist here as well. Identifying grammatical errors is a task the AI handles with precision, scanning your text for common mistakes and offering corrections. This feature acts as a preliminary check before you engage in more in-depth editing. ChatGPT also suggests stylistic improvements, proposing alternative phrasing or sentence structures that enhance clarity and readability. These suggestions are not just about fixing errors but also about refining your prose to ensure it flows smoothly and effectively communicates your ideas. By leveraging ChatGPT for editing and proofreading, you gain a second pair of eyes that complements your efforts, ensuring your writing reaches its full potential.

In this collaborative landscape, ChatGPT is more than a tool; it's a partner in your creative journey. Whether you're developing intricate plots or experimenting with new styles, its capabilities offer support and inspiration. The synergy between human creativity and AI extends beyond mere assistance, creating opportunities for exploration and growth in your writing endeavors.

## 6.2 LEVERAGING AI FOR MARKET RESEARCH

In the fast-paced world of market research, efficiency is everything. Imagine being able to sift through mountains of consumer feedback effortlessly, extracting valuable insights without breaking a sweat. With ChatGPT, automating data collection becomes a reality. This AI can efficiently pull insights from consumer reviews, identifying common themes and sentiments that might otherwise go unnoticed. By analyzing these reviews, you can discern patterns in customer satisfaction and areas for improvement, offering a clearer picture of consumer expectations. Additionally, compiling competitor analysis reports no longer requires hours of manual labor. ChatGPT can aggregate and organize information about competitors, highlighting their strengths and weaknesses. This automated approach allows you to stay informed about market dynamics and adjust your strategies accordingly.

Understanding market trends is vital for staying ahead. ChatGPT offers powerful techniques for trend analysis and forecasting, making it easier to navigate the ever-changing landscape. By analyzing social media sentiment, ChatGPT can gauge public opinion and identify emerging trends. This analysis involves scanning platforms for keywords and phrases associated with your industry, providing real-time insights into consumer attitudes. Predicting emerging consumer needs is another area where ChatGPT excels. By examining past data and current trends, the AI can forecast future demands, helping you anticipate market shifts before they occur. This foresight is invaluable for developing products and services that meet evolving consumer expectations, ensuring that your offerings remain relevant and competitive.

Creating detailed customer personas is an art, one that ChatGPT makes more precise through its data analysis capabilities. By segmenting audiences based on behavior patterns, ChatGPT identifies distinct consumer groups, each with unique preferences and needs. This segmentation allows you to tailor your marketing efforts more effectively, targeting messages that resonate with specific audiences. Developing narrative profiles for these target demographics further enhances your understanding of the customer. ChatGPT can construct these profiles by combining demographic data with behavioral insights, painting a vivid picture of your ideal customer. These profiles guide your marketing strategies, ensuring that your communication aligns with the values and interests of your audience.

In the competitive world of business, strategy is everything. ChatGPT's insights can refine and enhance your competitive strategies, giving you a distinct edge. Developing unique selling propositions becomes more straightforward with AI's analytical capabilities. By identifying what sets your product apart, ChatGPT helps craft a compelling narrative that highlights your strengths. Identifying gaps in market offerings is another area where ChatGPT proves invaluable. By analyzing industry trends and consumer feedback, the AI pinpoints unmet needs, offering opportunities for innovation. This insight allows you to capitalize on market voids, positioning your products as solutions to existing problems. With ChatGPT, you can navigate the complexities of market research with ease, using AI as a powerful ally in your strategic planning.

## 6.3 CONTINUOUS SKILL IMPROVEMENT STRATEGIES

In the landscape of AI, setting personal development goals becomes an anchor in your growth journey with ChatGPT. Establishing milestones for skill progression is not just about ticking boxes; it's about creating a roadmap for where you want to go with your AI capabilities. These milestones might include mastering specific prompt techniques or exploring new applications of AI in your work or hobbies. Tracking performance metrics and growth can offer valuable insights into your progress. This might involve assessing the complexity of tasks you can handle independently or noting improvements in efficiency when using AI tools. By monitoring these metrics, you gain a clearer picture of your development, allowing for targeted improvements and celebrating achieved goals.

Joining peer learning networks can amplify your growth in unexpected ways. These communities, whether online or in-person, provide a platform for shared learning and collective skill enhancement. Participating in AI-focused forums and discussions exposes you to diverse perspectives and techniques, broadening your understanding of what AI can achieve. Engaging in conversations with fellow learners can spark new ideas and challenge your thinking, pushing you beyond your comfort zone. Collaborating on group projects and challenges within these networks also enhances your practical skills. Working alongside peers on AI projects not only fosters teamwork but also offers opportunities to learn from others' experiences and expertise. These interactions create a dynamic learning environment where knowledge is shared freely and growth is a communal effort.

Online courses and tutorials are invaluable resources for furthering your skills and knowledge in using ChatGPT. Enrolling in specialized AI courses can deepen your understanding of complex topics and introduce you to the latest advancements in the field. These courses often provide structured learning paths, guiding you through the intricacies of AI step by step. Following expert-led tutorials and workshops can also enhance your learning experience. These sessions, often available on platforms like Coursera or Udemy, allow you to learn directly from industry leaders and gain insights into best practices. By investing time in these educational resources, you not only expand your knowledge but also stay current with emerging trends and techniques, ensuring that your skills remain relevant and competitive.

Reflecting on personal progress is a crucial component of continuous skill improvement. Regular self-reflection helps you evaluate your achievements and identify areas for further development. Journaling experiences and learnings can be a powerful tool in this process. By documenting your interactions with ChatGPT, you create a record of your growth, capturing both successes and challenges. This practice encourages introspection and provides a reference point for measuring progress. Seeking feedback from peers and mentors is another valuable strategy. Constructive feedback can offer new perspectives and highlight blind spots you might overlook. By embracing feedback with an open mind, you gain clarity on your strengths and areas for improvement, guiding your efforts towards meaningful growth.

As you refine your skills with ChatGPT and develop new strategies, remember that learning is a continuous process. Each interaction offers a lesson, each challenge an opportunity to grow. By setting clear goals, engaging with peers, utilizing educational resources, and reflecting on your progress, you cultivate an environment where continuous improvement thrives. In doing so, you not only enhance your proficiency with AI but also prepare yourself to tackle new challenges and explore innovative applications of ChatGPT. This journey of skill development is ongoing, leading you to new horizons and possibilities with every step. In the next chapter, we will explore how ethical considerations in AI usage can further enhance both personal and professional experiences, ensuring that technology is used responsibly and effectively.

**Title:** Make a Difference with Your Review

**Subtitle:** The Power of Giving

"Money can't buy happiness but giving it away can." - Freddie Mercury

When we give without asking for anything in return, it makes us feel happier. Let's spread that happiness together!

Would you help someone like you—curious about ChatGPT and Artificial Intelligence but unsure where to start?

My goal is to make ChatGPT and Artificial Intelligence easy and fun for everyone, but I can't do it alone. I need your help to reach more people.

Most people choose books because of reviews, so I'm asking you to help someone just starting their ChatGPT journey by leaving a review.

It's free, takes less than a minute, and your review could make a difference:

- It could help a small business grow.
- It could support an entrepreneur.
- It could give an employee meaningful work.
- It could even help someone change their life.

To make a difference, just click the link below and leave a review:

**LEAVE A BOOK REVIEW**

If you enjoy helping others, you're my kind of person. Thank you from the bottom of my heart!

-Dale Tunnell

# CHAPTER SEVEN

## ETHICAL USE OF AI

Imagine this: you're at a bustling train station, and each train represents a stream of data zooming past you. Some trains carry sensitive information, private details that could expose someone's identity if mishandled. In this digital age, data is constantly in motion, much like those trains, and the tracks they follow are the regulations ensuring they reach their destination safely. As you delve into the world of AI, understanding the importance of privacy and data security becomes imperative, just as a station master ensures the safe passage of trains. These principles form the backbone of ethical AI use, safeguarding both individual rights and organizational integrity.

Understanding data privacy regulations is crucial for anyone using AI. In the European Union, the General Data Protection Regulation (GDPR) sets a high standard for data protection, influencing how AI systems operate. GDPR mandates that data processing is lawful only if there's a legitimate reason, such as consent or contractual necessity. It emphasizes data minimization, meaning only essential data should be collected, and purpose limitation, ensuring data is used only for specified purposes. For AI users, this means implementing privacy-preserving technologies from the outset, integrating data protection into the design of AI systems. This approach is not merely a legal obligation but a commitment to respecting user rights, fostering trust in AI applications.

In the United States, the California Consumer Privacy Act (CCPA) complements GDPR by granting California residents specific rights over their personal data. It requires businesses to disclose data collection practices and allows individuals to opt out of data sales. AI users must navigate these regulations carefully, ensuring transparency and user control over personal information. This legal landscape may seem daunting, but it underscores the importance of responsible data handling. Adhering to these regulations not only avoids legal repercussions but also demonstrates a commitment to ethical AI use, aligning with societal expectations of privacy and security.

Implementing data security measures is vital for protecting user information in AI applications.

Encryption serves as a robust defense, converting data into a code to prevent unauthorized access. It's like locking sensitive documents in a secure vault, accessible only to those with the key. Regularly updating security protocols is equally important, protecting systems against evolving threats. This practice ensures that defenses remain strong, much like fortifying a castle to withstand modern weaponry. By prioritizing security, you protect not only your data but also the trust that users place in your AI systems. This proactive stance is essential in a world where data breaches can have significant consequences, both for individuals and organizations.

User consent and transparency are foundational to ethical AI use. Obtaining clear consent involves designing consent forms that are straightforward and understandable, ensuring users know what they're agreeing to. Imagine signing a contract where every clause is explained, leaving no room for confusion. This clarity extends to data usage policies, which should be communicated effectively, outlining how data is collected, used, and shared. Transparency builds trust, reassuring users that their data is handled responsibly. By maintaining open communication, you empower users to make informed decisions about their data, reinforcing the ethical standards that underpin AI applications.

Minimizing data collection is another critical aspect of ethical AI use, reducing privacy risks by collecting only what is necessary. This principle aligns with the notion of traveling light, carrying only essential items to avoid unnecessary burden. Implementing data anonymization techniques further protects privacy, stripping data of identifiable information to prevent tracing it back to individuals. This practice ensures that even if data is compromised, personal identities remain secure. By adopting these strategies, you not only comply with privacy regulations but also enhance the ethical integrity of your AI systems, contributing to a digital ecosystem that respects and protects user privacy.

**Interactive Element: Data Privacy Checklist**

- ❖ Familiarize yourself with GDPR and CCPA requirements for data processing.
- ❖ Implement encryption to safeguard user information.
- ❖ Design clear consent forms and communicate data usage policies effectively.
- ❖ Limit data collection to essential information only.
- ❖ Use data anonymization techniques to protect user identities.

These practices outline a framework for ethical AI use, guiding you in creating systems that prioritize user privacy and security. By integrating these principles into your AI applications, you contribute to a more trustworthy and responsible digital environment.

## 7.1 ENSURING BIAS-FREE AI INTERACTIONS

In today's increasingly AI-driven world, recognizing potential sources of bias within AI systems is crucial for fair and equitable outcomes. Bias in AI can stem from various factors, with training data being a primary culprit. When datasets used to train AI models lack diversity, they tend to mirror the existing societal biases embedded within them. For instance, if a facial recognition system is trained primarily on images of light-skinned individuals, it may perform poorly on darker skin tones, perpetuating inaccuracies and unfair treatment. Similarly, algorithmic design can introduce biases when developers inadvertently encode their own prejudices or assumptions into AI systems. This can result in biased decision-making processes that favor certain groups over others. As users and developers of AI, it's vital to scrutinize these elements to ensure the systems we rely on are as unbiased and fair as possible.

To address these challenges, adopting strategies for mitigating bias is essential. One effective approach is diversifying training datasets. By including a wide range of data that represents different demographics, regions, and perspectives, we can train AI systems to be more inclusive and accurate. This diversity helps AI understand and respond to various cultural nuances and contexts, reducing the risk of biased outputs. Regularly auditing AI outputs for bias is another critical strategy. By systematically reviewing AI decisions and outcomes, we can identify any patterns of bias that emerge and take corrective action. This ongoing vigilance ensures that AI systems remain fair and equitable, continually improving their performance over time. These proactive measures not only mitigate bias but also enhance the credibility and reliability of AI applications.

Promoting inclusive AI practices requires a commitment to diversity within AI development teams. Engaging diverse teams in AI development brings a variety of perspectives and experiences, enriching the design and functionality of AI systems. When team members come from different backgrounds, they can identify potential biases and suggest ways to address them during the development process. Incorporating feedback from underrepresented groups is equally important. By actively seeking input from those who may be affected by AI decisions, developers can ensure that their systems serve everyone fairly and justly. This feedback loop fosters a culture of inclusivity, where AI systems are designed with the needs and concerns of all users in mind. The collaboration between diverse teams and communities not only reduces bias but also drives innovation, resulting in AI technologies that reflect and respect the diversity of the world we live in.

Setting up robust monitoring and feedback mechanisms is vital for assessing and addressing bias in AI systems. Implementing real-time bias monitoring tools allows organizations to detect and respond to bias as it occurs, preventing it from influencing critical decisions. These tools can flag potentially biased outcomes, enabling swift intervention to correct and refine AI models. Encouraging user feedback for bias detection is another important component of this process. Users provide invaluable insights into how AI systems function in real-world settings and can highlight instances of bias that may not be apparent to developers. By actively soliciting and valuing this feedback, organizations can create a continuous improvement loop, where AI systems are regularly evaluated and adjusted to ensure fairness and accuracy. This iterative approach to monitoring and feedback not only helps maintain bias-free AI interactions but also builds trust and confidence among users.

Through careful examination of bias sources, strategic mitigation efforts, and inclusive practices, we can ensure that AI systems operate fairly and equitably. This approach requires a commitment to diversity, openness to feedback, and a willingness to adapt and improve. As we continue to integrate AI into various aspects of our lives, maintaining these principles will be key to developing systems that serve everyone with integrity and respect. This ongoing journey toward fairness in AI underscores the importance of vigilance and responsibility in harnessing the potential of technology for the greater good.

## 7.2 SUSTAINABLE AI PRACTICES

Artificial Intelligence is rapidly transforming our world, but it comes with environmental considerations that cannot be ignored. As AI systems grow more complex, so does their appetite for energy. Training large AI models can consume vast amounts of electricity. Picture the energy it takes to power a small town—that's what these systems might require. This energy use contributes significantly to carbon emissions, largely due to the data centers that support AI operations. These centers, housing countless servers, consume electricity at impressive rates, often relying on non-renewable energy sources. It's crucial to recognize the carbon footprint they leave and seek ways to mitigate it. Sustainability in AI is not just an add-on; it's a necessity for ensuring that technological advancement does not come at the expense of our planet.

To address these environmental impacts, we must adopt sustainable practices in AI development. One effective strategy involves optimizing algorithms for energy efficiency. By designing algorithms that require less computational power, we can reduce energy consumption without sacrificing performance. This optimization is akin to tuning a car engine for better fuel efficiency, ensuring that every watt of energy is used wisely. Additionally, utilizing renewable energy sources is a promising avenue. Solar panels and wind turbines can power data centers, cutting down on fossil fuel reliance. This shift not only reduces emissions but also aligns AI operations with broader sustainability goals. As more organizations embrace these changes, the cumulative effect can lead to substantial environmental benefits, making AI a greener technology.

Responsible resource usage is another key component of sustainable AI. By reducing data storage requirements, organizations can lessen their environmental impact. This reduction involves careful management of data, retaining only what's necessary and discarding redundant or outdated information. It's like cleaning out a cluttered garage, keeping only what's essential and useful. Implementing efficient data processing techniques further supports this goal. Techniques that streamline data handling reduce the computational burden, lowering energy use. These methods ensure that systems run smoothly and sustainably, minimizing their ecological footprint. Through thoughtful resource management, AI can evolve into a model of efficiency, showcasing how technology can advance without depleting the earth's resources.

Sustainability in AI is not just about individual practices; it's about fostering a culture that prioritizes green technology. Advocating for sustainability in AI policies is crucial for creating lasting change. This advocacy involves incorporating sustainability goals into AI projects from the outset, ensuring that environmental considerations are integral to the planning and execution phases. Supporting research in eco-friendly AI technologies is another vital step. By investing in innovative solutions that reduce environmental impact, we can drive the development of AI systems that are both powerful and sustainable. Encouraging policies that prioritize sustainability not only benefits the environment but also positions AI as a responsible and forward-thinking industry. Through collaboration and commitment, we can reshape how AI interacts with the environment, paving the way for a future where technology enhances rather than depletes our world.

**Textual Element: Resource List**

- Explore renewable energy sources like solar and wind for AI operations.
- Optimize algorithms for energy efficiency to reduce consumption.
- Implement data storage reduction strategies to minimize resource use.
- Support research initiatives focused on eco-friendly AI technologies.

## 7.3 BUILDING TRUST IN AI SYSTEMS

Trust in AI systems hinges on transparency, a cornerstone that ensures users feel confident and secure in the technology guiding parts of their lives. Transparency in AI means shedding light on the decision-making processes that occur within these complex systems. Imagine trying to understand a magic trick without seeing the magician's hands. Similar to this analogy, AI systems require clarity in how decisions are made.

This includes explaining the algorithms and data inputs that inform each outcome. By providing these insights, users can see how their interactions are processed, making the AI's actions more predictable and understandable. Clarifying how AI models are trained is equally crucial. Users should know the data sources and training methods employed, which helps in assessing the system's reliability and fairness. This openness demystifies AI, transforming it from a black box into a tool whose workings are as clear as a well-lit room.

Establishing accountability mechanisms in AI systems further bolsters trust. In any system where decisions have significant impacts, it's essential to identify who is responsible for those outcomes. This responsibility ensures that if errors occur, there are clear channels for addressing them. Implementing accountability frameworks means setting up processes that monitor and evaluate AI decisions continuously. These frameworks can include regular audits and assessments that ensure AI systems function as intended. By assigning accountability, organizations take ownership of their AI systems, ready to address any issues transparently and efficiently. This proactive stance not only mitigates risks but also builds user confidence, knowing that there is a structure in place to rectify mistakes and uphold standards.

Enhancing user understanding of AI systems is another pillar in building trust. Users need to feel confident in their interactions with AI, knowing that they fully grasp how the technology works. Educational initiatives aimed at improving AI literacy can play a pivotal role here. By demystifying AI concepts and jargon, these programs empower users to engage with AI more effectively. Clear communication of AI capabilities and limitations is also vital. Users should understand what AI can and cannot do, which helps set realistic expectations and prevents misunderstandings. When users are well-informed, they become more comfortable using AI systems, enhancing their overall experience and trust in the technology.

Fostering collaborative trust involves bringing together various stakeholders in the AI ecosystem. Engaging user communities in AI design ensures that the technology reflects the needs and concerns of those it serves. This collaboration can take many forms, from user feedback sessions to community-led design initiatives. By involving users in the development process, AI systems become more aligned with real-world needs and expectations. Building partnerships with ethical AI organizations further strengthens this trust. These partnerships can provide guidance, resources, and support in developing AI systems that adhere to ethical standards. Together, these collaborative efforts create a foundation of trust, where users feel confident that AI systems are designed with their best interests in mind.

In the grand tapestry of AI, trust is the thread that holds everything together. Transparency, accountability, user understanding, and collaboration weave a fabric that supports the responsible and ethical use of AI. By focusing on these elements, we ensure that AI systems serve humanity positively and effectively. As we look to the future, the need for trust will only grow more critical. AI's potential to transform industries and societies is immense, but this potential can only be realized if users trust the technology. In the next chapter, we will explore how AI can enhance human creativity, further broadening the horizons of what is possible when technology and humanity work hand in hand.

# CHAPTER EIGHT

## CASE STUDIES AND REAL-WORLD APPLICATIONS

Imagine a bustling newsroom where deadlines loom and the pressure to produce high-quality content is relentless. Editors scramble to curate articles, journalists juggle multiple stories, and creative blocks threaten productivity. Enter ChatGPT, an AI tool that has become a game-changer in content creation, transforming how publishing and media industries operate. By streamlining editorial workflows, ChatGPT frees up resources, allowing professionals to focus on crafting compelling narratives and engaging with their audience.

In the realm of publishing, ChatGPT excels by automating content curation and suggesting relevant topics. Picture an editor faced with a blank slate, unsure of what stories to pursue. ChatGPT analyzes current trends and audience interests, providing a curated list of topics that resonate with readers. This capability not only saves time but also ensures that content remains relevant and timely.

Furthermore, ChatGPT assists in generating and optimizing headlines, a crucial component of capturing audience attention. By analyzing linguistic patterns and engagement metrics, ChatGPT suggests headlines that are both catchy and SEO-friendly, driving traffic and increasing visibility. The result is a more efficient editorial process where creativity can flourish without the burden of routine tasks.

Consider a digital magazine that embraced ChatGPT to enhance its content output and quality. Before integrating AI, the team struggled with tight deadlines and limited resources. ChatGPT became an invaluable tool, assisting in article drafting and enabling rapid fact-checking and editing. The AI's ability to generate coherent drafts from simple prompts allowed writers to focus on refining their stories rather than starting from scratch. Additionally, ChatGPT's editing capabilities ensured that articles were polished and free from grammatical errors, providing quick feedback for improvements. This transformation not only increased the magazine's content production but also elevated its quality, cementing its reputation as a reliable source of information.

Audience engagement is paramount in the media industry, and ChatGPT plays a pivotal role in personalizing content to enhance interaction. By conducting an AI-driven audience analysis, ChatGPT identifies reader preferences and recommends content tailored to their interests. This personalization fosters a deeper connection between the publication and its audience, increasing loyalty and satisfaction. Moreover, ChatGPT integrates real-time feedback into content adjustments, allowing publishers to adapt their strategies based on audience reactions. This dynamic approach ensures that content remains relevant and engaging, encouraging readers to return for more.

Authors, too, have found an ally in ChatGPT, particularly when faced with writer's block. The AI assists in developing plot outlines and exploring character development, providing fresh perspectives that inspire creativity. Imagine an author struggling to flesh out a character's backstory. By feeding ChatGPT a few details, the AI generates dialogues and scenarios that breathe life into the character, sparking new ideas and directions for the narrative. This collaboration not only breaks through creative barriers but also expands the author's storytelling toolkit, offering new avenues for exploration.

### Reflective Exercise: Content Creation with ChatGPT

Reflect on a recent writing project. Identify a moment when you faced a creative block or uncertainty in your narrative. Consider how ChatGPT could assist in this scenario. Draft a prompt that describes your challenge and review the AI's suggestions. Analyze how these insights could enhance your writing, offering new directions or solutions. This exercise encourages you to leverage ChatGPT as a creative partner, transforming obstacles into opportunities for innovation.

In the ever-evolving landscape of media and content creation, ChatGPT stands out as a versatile and indispensable tool. By enhancing editorial workflows, improving audience engagement, and supporting creative writing, ChatGPT empowers professionals to deliver high-quality content with efficiency and flair. Its impact extends beyond automation, fostering a collaborative environment where human creativity and AI capabilities intersect, driving the industry forward.

## 8.1 TRANSFORMING MARKETING STRATEGIES WITH AI

In the fast-paced world of marketing, the ability to adapt and innovate is crucial. ChatGPT emerges as a powerful tool that revolutionizes how businesses approach their marketing campaigns. At the core of this transformation is the AI's capacity for keyword optimization, a vital component for enhancing SEO. By analyzing vast datasets, ChatGPT identifies the most effective keywords that align with current trends, ensuring your content ranks higher in search engine results. This optimization goes beyond mere visibility, targeting the right audience with precision. Personalized ad copy generation is another area where ChatGPT excels. Utilizing user data, the AI crafts tailored messages that resonate with individual consumers. Imagine creating an ad campaign where each piece of content speaks directly to the viewer's preferences and behaviors. The result is a boost in engagement and conversion rates, as consumers feel understood and valued.

Consider an e-commerce platform that capitalized on ChatGPT's capabilities to enhance customer satisfaction and sales. The platform integrated chatbots powered by ChatGPT to offer 24/7 customer support. These chatbots provided instant responses to queries, guiding users through purchasing decisions and resolving issues efficiently. The constant availability of support not only reduced cart abandonment rates but also fostered customer loyalty. Moreover, the platform implemented automated product recommendation systems using ChatGPT's insights. By analyzing user behavior and preferences, the AI suggested products that matched individual tastes, leading to increased sales and customer delight. This case demonstrates how leveraging AI transforms e-commerce operations, creating a seamless and personalized shopping experience that keeps customers returning for more.

Understanding market trends and consumer preferences is a cornerstone of strategic decision-making. ChatGPT aids this process through its adeptness in sentiment analysis on social media platforms. By scanning posts, comments, and reviews, the AI gauges public opinion and identifies emerging trends. This analysis provides invaluable insights into consumer sentiment, allowing businesses to adjust their strategies proactively. Competitive analysis is another area where ChatGPT shines. By dissecting competitors' tactics and analyzing market movements, the AI offers a comprehensive view of the competitive landscape. These insights enable companies to refine their approaches, anticipate market shifts, and capitalize on opportunities, all while staying a step ahead of their rivals.

Brand storytelling is an art that captures the essence of a company, and ChatGPT plays a pivotal role in crafting narratives that resonate with audiences. By generating unique brand stories, the AI helps businesses communicate their values and missions effectively. These narratives go beyond traditional advertising, creating emotional connections with consumers that foster brand loyalty. Furthermore, ChatGPT supports the development of interactive marketing campaigns. By generating AI-driven content, brands can engage audiences in novel ways, such as interactive quizzes or personalized video messages. These campaigns invite participation, making consumers feel like active contributors to the brand's story. The result is heightened engagement and a stronger bond between the brand and its audience.

### Interactive Element: Marketing Strategy Checklist

Reflect on your current marketing strategies. Use ChatGPT to analyze your keyword optimization and ad copy effectiveness. Identify areas for improvement and draft a personalized ad campaign using AI insights. Consider how ChatGPT can enhance your brand storytelling and engage your audience in new ways. This exercise encourages you to explore AI's potential in transforming your marketing efforts, fostering innovation and growth.

## 8.2 EDUCATIONAL USES OF CHATGPT

In today's rapidly evolving educational landscape, personalized learning has taken center stage, and ChatGPT is proving to be an invaluable ally. Imagine a classroom where each student progresses at their own pace, guided by lessons that cater to their unique strengths and weaknesses. ChatGPT facilitates this by creating adaptive learning modules tailored to individual needs. Whether a student excels in math but struggles with language arts, ChatGPT adjusts the content to provide additional practice where it's needed and advanced challenges where appropriate. This adaptability ensures that no student is left behind, allowing educators to focus their attention on guiding and inspiring rather than merely lecturing. Real-time feedback and progress tracking are also possible with ChatGPT. Students can receive immediate responses to assignments, highlighting areas of improvement and encouraging self-reflection. This instant interaction not only accelerates learning but also fosters a sense of responsibility and independence in learners, preparing them for lifelong educational pursuits.

Consider an online learning platform that has integrated ChatGPT to revolutionize its approach to student engagement and comprehension. This platform offers interactive Q&A sessions where students can pose questions and receive detailed explanations anytime, anywhere. This accessibility breaks down the barriers of traditional classroom settings, providing support outside regular school hours. ChatGPT also automates the grading process, offering comprehensive feedback on assignments. By analyzing elements like coherence, structure, and clarity, the AI provides constructive critique that helps students refine their work. This automated process saves educators time, allowing them to dedicate more energy to developing curriculum and supporting students individually. The result is a dynamic learning environment where students feel supported and motivated to excel.

Educators themselves benefit significantly from ChatGPT, as it alleviates the burden of administrative tasks and enriches teaching methods. Lesson planning, often a time-consuming endeavor, becomes more efficient with AI assistance. ChatGPT suggests engaging activities and materials that align with educational standards, ensuring that lessons are both relevant and captivating. Additionally, the AI aids in resource creation, generating content that caters to diverse learning styles. Whether it's visual aids for visual learners or interactive exercises for kinesthetic learners, ChatGPT helps educators create a more inclusive classroom environment. This support allows teachers to focus on what they do best: inspiring and guiding the next generation.

Fostering creativity in students is another area where ChatGPT shines. Through innovative assignments, students are encouraged to explore their imaginations and express themselves in new ways. Creative writing prompts generated by ChatGPT can serve as a springboard for students' ideas, helping them overcome the fear of a blank page. These prompts might include intriguing scenarios or character dilemmas that ignite curiosity and drive narrative exploration. Collaborative projects also benefit from AI involvement. With ChatGPT as a partner, students can experiment with different roles and perspectives, enhancing their teamwork and problem-solving skills. By encouraging students to think creatively and work collaboratively, ChatGPT helps cultivate essential skills that are valuable both in and out of the classroom.

### Interactive Element: Creative Assignment Exercise

Reflect on an assignment you've given or received that involved creative writing or problem-solving. Use ChatGPT to generate a prompt that could be used to inspire creativity in a similar context.

Share this prompt with a peer or student and discuss how it could be expanded into a complete project. Consider how AI can enhance the learning experience by providing diverse perspectives and fostering innovative thinking.

As education continues to evolve, the potential for ChatGPT to enrich learning experiences is vast. Its ability to personalize content, support educators, and foster creativity makes it a powerful tool in the modern classroom. By integrating AI into education, we open new pathways for learning, ensuring that students are equipped with the skills and knowledge to thrive in an ever-changing world.

## 8.3 AI IN FREELANCING: A SUCCESS STORY

In the competitive world of freelancing, efficiency and adaptability are key. Freelancers often juggle multiple projects and deadlines, requiring a streamlined approach to manage their workload effectively. ChatGPT emerges as a vital tool in this landscape, enhancing workflow efficiency and boosting project output. For instance, freelancers can automate client communication and follow-ups, saving significant time and effort. Instead of manually crafting each email or message, they can use ChatGPT to generate personalized templates that maintain a professional tone. This automation ensures that client relationships remain strong, even as the freelancer manages a growing portfolio. Additionally, AI-driven scheduling tools assist freelancers in organizing their time more effectively. With ChatGPT's help, they can set reminders, prioritize tasks, and allocate time slots for different projects, ensuring that deadlines are met and quality is maintained.

Take the case of a freelance content creator who transformed their business using ChatGPT. Initially, this freelancer struggled to keep up with the demands of writing, editing, and managing client interactions. By integrating AI capabilities, they expanded their service offerings, providing clients with data-driven insights and content strategies. ChatGPT's ability to quickly generate content drafts and suggestions allowed the freelancer to enhance content quality while reducing turnaround time. The increased efficiency not only boosted the freelancer's income but also significantly improved client satisfaction. Clients appreciated the swift delivery and high-quality work, leading to repeat business and positive referrals. This case illustrates how embracing AI can lead to substantial growth in a freelancing career, opening doors to new opportunities and elevating professional standards.

Networking and client acquisition are crucial for freelancers seeking to grow their business. ChatGPT plays a pivotal role in this process by crafting personalized pitches tailored to potential clients. With AI assistance, freelancers can analyze client needs and preferences, tailoring their proposals to address these specific points. This personalized approach increases the chances of securing new projects, as clients feel understood and valued. Furthermore, ChatGPT aids in identifying potential clients through AI-driven market analysis. By examining industry trends and competitor strategies, freelancers can pinpoint opportunities and target clients who align with their skills and expertise. This targeted approach not only saves time but also enhances the effectiveness of networking efforts, ensuring that freelancers connect with the right people in their field.

Diversifying skills and services is another area where ChatGPT proves invaluable for freelancers. The AI offers tutorials and resources to help freelancers learn new skills, expanding their expertise and service offerings. Whether it's mastering a new software tool or exploring a different aspect of their industry, ChatGPT provides guidance and support. Additionally, freelancers can leverage AI to offer enhanced services such as data analysis and visualization. By incorporating these capabilities, they can provide clients with comprehensive solutions that go beyond traditional freelancing roles. This diversification not only increases the freelancer's value to clients but also opens up new revenue streams, ensuring long-term success and stability in an ever-changing market.

This chapter highlights how ChatGPT revolutionizes freelancing by improving workflow efficiency, enhancing content quality, and expanding service offerings. The integration of AI into freelancing empowers individuals to manage their businesses more effectively, connect with clients strategically, and continuously develop their skills. As we move forward, the next chapter will explore how AI impacts broader business strategies and operations, offering insights into its transformative potential across various industries.

# CHAPTER NINE

## ENHANCING HUMAN CREATIVITY WITH AI

Imagine a painter standing before a blank canvas, brush in hand, pondering the possibilities of color and form. Now imagine that alongside this artist stands an invisible assistant, an AI partner whispering suggestions, offering insights, and helping shape the creative process. This is not a scene from a science fiction novel but a reality that many creatives are beginning to explore. AI tools like ChatGPT are emerging as valuable collaborators, transforming the traditional creative process by introducing new ways to brainstorm, refine, and innovate. As you delve into this chapter, we'll explore how AI can serve as a creative partner, enhancing your artistic capabilities and opening doors to uncharted territories.

In the realm of creative collaboration, AI has redefined what it means to work alongside a partner. No longer confined to human colleagues, creatives now have the option to engage with AI as an adaptive collaborator. Imagine you're in a brainstorming session, searching for that elusive spark of inspiration. ChatGPT can generate a multitude of concepts, each with its own twist and potential. Whether you're crafting a marketing campaign or developing a new product, this AI-generated input can broaden your perspective, offering ideas that might not have occurred to you otherwise. Moreover, AI introduces adaptive feedback loops, refining creative outputs in real-time. As you provide input, ChatGPT adjusts its responses, creating a dynamic exchange that evolves with your needs. This adaptability enhances the creative process, allowing for more fluid and responsive collaboration.

Beyond ideation, ChatGPT plays a significant role in streamlining creative workflows. Consider the myriad tasks involved in any creative project, from initial brainstorming to final execution. These tasks often include repetitive elements, such as drafting initial versions or organizing project elements. ChatGPT can automate these repetitive tasks, freeing you to focus on higher-level creative decisions. Imagine working on a script or article and letting ChatGPT handle the first draft. This automation not only saves time but also helps overcome creative inertia, that feeling of being stuck at the start of a project. By generating initial content, ChatGPT provides a foundation upon which you can build, allowing your creative energy to be directed towards refining and enhancing rather than starting from scratch.

One of the most exciting aspects of partnering with AI is its ability to introduce diverse perspectives. As a tool, ChatGPT can explore themes and ideas that might be unconventional or outside your usual scope. For instance, it can suggest incorporating multicultural influences into a design project, offering insights drawn from a broad spectrum of cultural contexts. This ability to pull from a wide array of sources enriches your work, adding layers of depth and novelty. By engaging with these diverse perspectives, you can expand your creative horizons, breaking free from traditional constraints and discovering new avenues for expression. This exploration is particularly valuable in today's globalized world, where understanding and integrating diverse viewpoints can lead to more inclusive and impactful creative work.

The synergy between human intuition and AI logic represents a powerful fusion in the creative process. While AI excels at structuring complex narratives and analyzing patterns, it lacks the intuitive understanding that human creators bring to their work. This is where your unique insights come into play, blending seamlessly with AI's analytical capabilities. You might use AI logic to outline a complex story or design, creating a cohesive framework that guides your project. Then, your intuition and creativity fill in the gaps, adding emotion, nuance, and depth that AI alone cannot achieve. This collaboration results in innovative designs and narratives that are both well-structured and rich in human touch. By balancing these elements, you harness the full potential of AI as a creative partner, enhancing your work while preserving the essence of your artistic vision.

**Interactive Element: Creative Prompt Exploration**

Consider a project you're currently working on or would like to start. Use ChatGPT to generate a list of creative prompts related to your project. Reflect on these prompts and choose one that resonates with you. Spend some time exploring the ideas it sparks, considering how AI's suggestions might influence your approach. This exercise encourages you to engage actively with ChatGPT, discovering new perspectives and possibilities in your creative endeavors.

## 9.1 COLLABORATIVE PROJECTS WITH CHATGPT

When we consider the possibilities of interdisciplinary projects, the integration of AI like ChatGPT becomes a game-changer. Picture a team of architects, engineers, and artists working together on a new urban development project. ChatGPT steps in as an invaluable mediator, weaving together the technical aspects with artistic flair. It facilitates communication across these diverse fields, ensuring that each team member—whether focused on structural integrity or aesthetic appeal—can contribute effectively. This integration helps create projects that are not only functional but also visually inspiring. By acting as a bridge, ChatGPT enables seamless exchanges of ideas, turning complexities into cohesive, innovative outcomes.

In the dynamics of teamwork, AI serves as a silent yet powerful ally. Creative differences, often a hurdle in collaborative settings, find resolution with AI-generated suggestions that offer compromise without sacrificing creativity. It's like having a neutral party that mediates discussions, proposing solutions that satisfy all parties involved. Additionally, AI encourages equal participation. In group scenarios, some voices might dominate while others go unheard. ChatGPT can moderate these interactions, ensuring that everyone's input is considered, fostering a balanced and inclusive environment. This equal footing encourages team members to contribute freely, knowing that their ideas will be acknowledged and valued.

Imagine a live brainstorming session, the room buzzing with energy, ideas flying fast and free. ChatGPT can be part of this dynamic exchange, offering real-time input that sparks spontaneous creativity. Whether it's a design firm brainstorming a new campaign or a theater group developing a script, AI's ability to contribute live can lead to moments of unexpected brilliance. It acts as a catalyst, prompting team members to think outside the box. In performing arts, for instance, ChatGPT can assist in improvisation, suggesting plot twists or character developments that might not have been considered otherwise. This spontaneity enriches the creative process, leading to innovations that surprise and delight both creators and audiences.

Feedback and iteration are crucial to refining any creative endeavor, and ChatGPT plays a significant role in facilitating these cycles. Rapid prototyping becomes more efficient with AI feedback, allowing teams to test and refine concepts quickly. Consider a product development team using ChatGPT to analyze initial prototypes. The AI provides insights that help identify potential improvements, guiding the team through iterative design processes. This continuous input ensures that each cycle of iteration builds upon the last, leading to a polished final product. By incorporating ChatGPT into these feedback loops, teams benefit from a steady stream of valuable insights, accelerating the journey from concept to completion.

**Interactive Element: Team Dynamics Reflection**

Reflect on a recent team project you've been involved in. Consider how ChatGPT could have facilitated communication and collaboration. Identify specific instances where AI might have helped mediate differences or encouraged participation. Jot down your reflections and discuss them with your team, exploring how AI can enhance future collaborations. This exercise encourages you to think critically about the role of AI in team dynamics, fostering a more cohesive and innovative approach to collaborative projects.

## 9.2 EXPANDING ARTISTIC BOUNDARIES

Art has always been about breaking new ground, challenging norms, and pushing the limits of what is possible. With the advent of AI, artists are finding new ways to expand the boundaries of traditional art forms. Imagine a textile designer who uses ChatGPT to generate intricate patterns that blend ancient techniques with modern aesthetics. By inputting different themes and styles, the designer receives unique pattern suggestions that might never have been conceived through traditional methods alone. This AI-driven innovation allows artists to explore new color palettes, combining hues in unexpected ways that evoke emotion and intrigue. The result is a fresh take on textile design that captivates and inspires, offering endless possibilities for creativity.

The intersection of art and technology is a fertile ground for innovation, and AI is playing a pivotal role in this cross-pollination. Consider digital installations that incorporate AI interactivity, where art responds to the viewer's presence and actions, creating a dynamic and immersive experience. These installations might include visual and auditory elements that change and evolve, engaging the audience in a dialogue with the art itself. Similarly, augmented reality experiences can be enriched with AI-driven narratives, allowing users to interact with virtual worlds layered over the real one. By merging digital technology with artistic expression, creators can craft experiences that transcend traditional boundaries, inviting viewers to participate in the art in ways that were previously unimaginable.

AI is not just transforming existing art forms; it's also inspiring the creation of entirely new genres and styles. Imagine a musician who uses AI to blend classical compositions with digital soundscapes, creating a fusion that is both familiar and novel. AI-influenced music composition techniques can introduce complex rhythms and harmonies, offering musicians a playground of endless possibilities. This fusion of classical and digital art forms doesn't just create new sounds; it redefines what music can be, pushing artists to explore uncharted territories. In visual arts, AI can help artists experiment with styles that combine traditional painting techniques with digital manipulation, resulting in works that challenge the viewer's perception and understanding of art.

Embracing experimental art with AI as a co-creator encourages artists to venture beyond the conventional. Algorithmic art and generative design allow creators to use AI algorithms to generate artworks that are both surprising and thought-provoking. These pieces might be created through processes that involve chance and randomness, inviting artists to relinquish some control and allow the machine to contribute to the creative process. Similarly, AI-driven performance art pieces can incorporate real-time data and interactions, creating performances that evolve and change with each iteration. These experimental approaches invite artists to question the nature of creativity and authorship, embracing AI as a collaborator that enhances, rather than diminishes, the human touch.

Exploring these boundaries requires an openness to new ideas and a willingness to experiment. Artists who engage with AI find themselves on the cutting edge of creativity, where traditional techniques meet modern technology. By pushing the limits of what is possible, they redefine art for a new generation, opening doors to forms and expressions that challenge and inspire. As technology continues to evolve, the potential for artistic innovation grows, inviting artists to explore, create, and imagine in ways that are as boundless as their creativity itself. The journey of art is ever-evolving, a testament to the relentless pursuit of expression and the transformative power of technology in the hands of visionary creators.

## 9.3 FOSTERING INNOVATION THROUGH AI

In the realm of creativity and innovation, ChatGPT emerges as a catalyst that propels ideas into action. Imagine standing at the crossroads of the creative industries, where untapped opportunities lie just beyond the horizon. ChatGPT can illuminate these paths by providing fresh insights that might otherwise remain hidden. Whether you're a filmmaker exploring a new narrative style or an entrepreneur seeking novel ways to engage audiences, AI's capacity to analyze trends and generate creative solutions can unveil possibilities that were previously out of reach. By processing vast amounts of data and identifying patterns, ChatGPT helps you pinpoint opportunities ripe for exploration, offering a strategic advantage in a rapidly evolving landscape.

When developing new concepts, the ability to rapidly generate and iterate on ideas is invaluable. ChatGPT accelerates this process by assisting in the creation of concept sketches and scenarios that form the backbone of strategic visioning. Consider a designer working on a new product. ChatGPT can produce a range of initial sketches, each offering a different take on the design brief. This rapid generation not only speeds up the prototyping phase but also allows for a broader exploration of possibilities. In strategic planning, AI-assisted scenario planning enables businesses to visualize potential futures and prepare for various outcomes. By simulating different scenarios, organizations can identify the most promising paths forward, ensuring that they are ready to adapt to changing circumstances.

Innovation doesn't stop at the concept stage; it permeates business models and entrepreneurial ventures. ChatGPT drives this innovation by helping businesses develop unique value propositions and refine their market positioning strategies. For a startup entering a competitive market, establishing a distinct identity is crucial. AI can analyze market trends and consumer preferences, offering insights that inform the development of a compelling value proposition. Additionally, AI-driven market positioning strategies leverage data to pinpoint niches that competitors may have overlooked. By aligning their offerings with customer needs, businesses can carve out a niche that sets them apart, ensuring long-term success in their chosen markets.

Cultivating a culture of innovation within organizations and communities requires more than just tools; it demands a shift in mindset. ChatGPT supports this shift by encouraging risk-taking and experimentation, essential components of any innovative culture. By providing a safe space for exploration, AI allows individuals and teams to test new ideas without fear of failure. This environment fosters creativity and encourages bold thinking, paving the way for breakthroughs that might otherwise remain undiscovered. Furthermore, AI facilitates innovation workshops, where participants can engage with AI tools to brainstorm and refine ideas. These workshops create a collaborative atmosphere, bringing together diverse perspectives and expertise to solve complex problems. Through these interactions, participants develop a deeper understanding of how AI can augment their creative processes, leading to a more innovative and agile organization.

As we conclude this chapter, remember that AI, particularly ChatGPT, is not just a tool but a partner in the creative process. It offers new ways to think about problems, explore solutions, and implement innovative ideas. By integrating AI into your workflows, you unlock a world of possibilities, where creativity and technology intersect to drive progress and change.

# CHAPTER TEN

## BUILDING A SUPPORTIVE LEARNING ENVIRONMENT

Imagine stepping into a vibrant market bustling with voices, each stall offering a unique taste of culture and knowledge. This is akin to joining AI enthusiast communities, where diverse perspectives converge, creating a rich tapestry of learning and collaboration. For beginners venturing into the world of ChatGPT and AI, these communities provide a supportive environment to explore, question, and grow.

Finding the right community can be as rewarding as discovering a new favorite book series. Start by exploring online forums like Reddit's AI subreddit, a hub for AI news, research papers, and tutorials. This platform is a treasure trove of information, where enthusiasts and experts share insights and engage in lively discussions. Similarly, Discord hosts numerous AI-focused servers, where real-time conversations and Q&A sessions unfold. These spaces encourage questions and foster connections, making them ideal for those who prefer interactive learning. For those who enjoy face-to-face interaction, participating in local AI meetups and events can be invaluable. Platforms like AI Meetup organize gatherings where enthusiasts share experiences and collaborate on projects, creating opportunities for networking and hands-on learning.

Engaging with these communities offers a multitude of benefits. One of the most significant is access to diverse perspectives and experiences. Each member brings a unique background, contributing to a richer understanding of AI's potential and challenges. This diversity is a wellspring of creativity and innovation, sparking ideas and solutions that a single viewpoint might overlook. Moreover, these communities are fertile grounds for collaboration on projects. Whether it's developing a new application or conducting research, working alongside others enhances both the learning experience and the quality of the outcomes. Knowledge sharing is another key advantage, as members exchange insights, resources, and best practices. This collaborative spirit accelerates learning, enabling members to tackle complex topics with greater confidence.

Active participation is crucial to reaping the full benefits of these communities. Begin by initiating discussions and asking questions. Curiosity is your greatest asset, and engaging with others through thoughtful inquiries not only enriches your understanding but also encourages others to share their knowledge. Consider sharing your personal experiences and insights as well. Whether you've discovered a new use case for ChatGPT or faced challenges in your AI journey, your contributions can offer valuable lessons to others. This reciprocal exchange of ideas fosters a sense of community and belonging, empowering you to grow alongside your peers.

Navigating community dynamics requires tact and open-mindedness. Disagreements are inevitable in diverse groups, but they can be constructive if handled with respect. Encourage respectful and constructive feedback, focusing on the merits of ideas rather than personal differences. This approach promotes a culture of inclusivity and diversity, where all voices are valued and heard. Inclusivity is not just about welcoming different perspectives; it's about creating an environment where everyone feels empowered to contribute. By embracing diverse viewpoints and championing respectful dialogue, you help build a community that thrives on collaboration and mutual respect.

**Reflective Exercise: Community Engagement Checklist**

- ❖ Identify at least three AI communities that align with your interests.
- ❖ Participate in one online forum discussion by asking a question or sharing an insight.
- ❖ Attend a local AI meetup or virtual event to network with fellow enthusiasts.
- ❖ Reflect on your experiences and consider how community engagement has enriched your understanding of AI.

By thoughtfully engaging with AI communities, you cultivate a supportive network that enriches your learning journey, providing you with the tools and insights needed to thrive in the rapidly evolving landscape of artificial intelligence.

## 10.1 UTILIZING ONLINE RESOURCES AND TUTORIALS

Navigating the vast sea of online resources can feel overwhelming, yet finding the right materials is key to effective learning in AI. Begin by evaluating the credibility of online courses, as not all are created equal. Look for courses on reputable platforms such as Coursera, edX, or DataCamp, where the content is often developed by industry experts and universities. Reviews and ratings can also provide insight into a course's quality and relevance. It's important to select tutorials that match your skill level and interests; starting with beginner-friendly resources can build a solid foundation before progressing to more advanced topics. Whether you're interested in machine learning, natural language processing, or data science, aligning your resources with your goals ensures that you remain engaged and motivated.

Creating a structured learning path is like crafting a roadmap for your educational journey, guiding you systematically through the complexities of AI. Begin by mapping out a learning timeline, considering how much time you can realistically dedicate each week to study and practice. Break down your learning objectives into specific milestones, such as completing a course module or mastering a particular concept. This approach not only organizes your learning but also provides a sense of accomplishment as you achieve each milestone. Keeping track of your progress helps maintain momentum and adapt your plan as needed, ensuring that your learning remains focused and efficient. Structure and organization transform a daunting learning process into a series of achievable steps.

Deciding between free and paid content involves weighing the pros and cons of each option. Free resources, such as YouTube tutorials and open-access articles, are abundant and often provide a good starting point for beginners. However, they may lack the depth and continuity of structured courses. Paid content, on the other hand, often offers comprehensive curricula, complete with assignments and certifications that can enhance your resume. When choosing paid options, consider whether the investment aligns with your learning goals and budget. Balancing both types of content allows you to maximize free resources while investing in paid courses that offer in-depth knowledge and guided learning. Thoughtful selection lets you tailor your learning experience to fit your personal needs.

Interactive content plays a pivotal role in making learning more engaging and effective. Participating in live webinars and Q&A sessions connects you with instructors and peers in real-time, allowing for immediate feedback and deeper understanding. These interactive elements transform passive learning into an active dialogue, where questions can be asked and concepts clarified on the spot. Interactive coding platforms, like Codecademy or Kaggle, offer hands-on practice with real-world datasets, reinforcing theoretical knowledge through practical application. Engaging with these platforms helps cement your understanding of AI concepts and develop problem-solving skills. This hands-on approach, where you can experiment and learn from trial and error, is invaluable in mastering the intricacies of AI.

## 10.2 PEER SUPPORT AND MENTORSHIP OPPORTUNITIES

Connecting with peers who share your interest in AI can be an invaluable part of your learning experience. These connections often begin in study groups found on online platforms. Joining such groups opens up a space where you can share knowledge, tackle challenges collectively, and motivate each other. Websites like Meetup or specific groups on platforms like Facebook can be great starting points. Within these communities, you can find accountability partners—fellow learners who meet regularly to check in on each other's progress. This mutual support system helps maintain focus and dedication, providing a sense of camaraderie that makes the learning process less daunting.

Mentorship is another powerful tool in accelerating your AI learning. Finding a mentor might seem daunting, but it's often about recognizing the potential mentors already present in your communities and networks. Look for individuals who have navigated the path you're on, those who can offer guidance and perspective from their experiences. Once you identify potential mentors, approach them with clear goals and expectations. It's important to communicate what you hope to achieve and how they can assist you. A well-defined mentorship relationship is mutually beneficial, offering structure and focus that significantly enhances learning and skill development.

Building a network of mentors and advisors involves strategic engagement and outreach. Networking events, both virtual and in-person, are prime opportunities to meet potential mentors. Attend AI conferences, webinars, and workshops where thought leaders gather and share insights. Engaging with professionals on platforms like LinkedIn can also be fruitful. Start by connecting with individuals whose work aligns with your interests, and initiate conversations that demonstrate your enthusiasm and willingness to learn. This proactive approach helps establish meaningful connections that can evolve into mentorships.

Giving back by mentoring others reinforces your own learning and contributes to community growth. Offering guidance to beginners in forums or study groups allows you to share your knowledge, which helps solidify your understanding. Hosting workshops or webinars to share expertise is another way to give back. This not only helps others but also positions you as a knowledgeable resource within the community. The act of mentoring fosters a cycle of learning and teaching, enhancing your skills and knowledge while supporting others on their AI learning paths.

### Interactive Element: Mentorship Reflection Exercise

Reflect on your current AI learning journey. Identify one area where you could benefit from mentorship and outline the steps you will take to find a mentor. Consider attending an upcoming networking event or reaching out to a professional on LinkedIn. Additionally, think about how you can offer your skills and knowledge to others, perhaps by answering questions in an online forum or organizing a small knowledge-sharing session.

## 10.3 ENGAGING IN CONTINUOUS LEARNING JOURNEYS

In the swiftly evolving landscape of artificial intelligence, adopting a mindset of lifelong learning becomes not just beneficial, but necessary. The digital world moves at a pace where yesterday's cutting-edge technology can become tomorrow's standard. Setting personal learning goals is a fundamental step in this ongoing process. Consider what you want to achieve in your AI exploration. Perhaps you're looking to deepen your understanding of machine learning, or maybe you're eager to apply AI tools like ChatGPT to creative projects. Regularly take a moment to reflect on your progress and achievements. This practice not only highlights areas where you've grown but also helps identify new directions for your learning. Reflection is key to maintaining motivation and ensuring your efforts align with your broader aspirations.

Technology itself offers tools to support this continuous learning mindset. Emerging learning technologies are transforming how we acquire new skills. AI-driven learning tools can integrate seamlessly into your daily routines, providing personalized feedback and adaptive learning paths that cater to your unique needs. Platforms like AI-driven apps for language learning, or adaptive coding tutorials, adjust the difficulty and focus based on your performance. This personalized approach ensures that learning remains challenging yet achievable, keeping you engaged and progressing at a comfortable pace. These tools can become part of your everyday routine, turning daily activities into opportunities for growth and development.

Staying informed about the latest trends and developments in AI is crucial to remaining relevant in a field known for rapid advancement. Following AI research publications and journals is a great way to stay updated. These sources offer insights into the latest breakthroughs and innovations, often providing a glimpse into the future of the technology. Subscribing to AI-focused newsletters and podcasts can also be beneficial. They deliver curated content directly to your inbox or ears, ensuring you receive the most relevant information without having to sift through vast amounts of material. These resources keep you connected to the pulse of AI advancement, offering context and understanding that enrich your learning.

A personal learning plan is your roadmap to achieving your educational and career goals. Begin by identifying skill gaps and areas where improvement is needed. This might involve assessing your current competencies against industry standards or future job requirements. Once you've pinpointed these areas, integrate learning activities into your daily schedules. Whether it's setting aside time each week for an online course, attending webinars, or dedicating a few minutes each day to reading AI news, these activities should fit seamlessly into your life. A well-structured learning plan offers clear direction and purpose, providing a framework that supports and enhances your growth.

As we draw this chapter to a close, remember that continuous learning allows you to adapt and thrive in the fast-paced world of AI. The next chapter will explore how to apply these skills practically, using AI to enhance creativity and problem-solving in everyday scenarios.

# CHAPTER ELEVEN

## ADDRESSING AI LIMITATIONS

Imagine you're at a bustling farmers market. Each stall represents a different culture, and every vendor speaks a unique dialect. You wander through, armed only with a translator app that struggles to grasp the nuances of local expressions and cultural references. This scenario mirrors some of the challenges AI faces today. While powerful, AI models like ChatGPT rely heavily on data for training and often struggle to fully understand the complexity of human language and context. As you explore the capabilities of AI, it's essential to recognize its limitations to better guide its use in your tasks.

Artificial intelligence, at its core, is data-driven. This dependency means AI models require vast amounts of high-quality data to learn and make decisions. Without this data, AI struggles to generate novel insights or adapt to new situations. Think of AI as a student who can only learn from the textbooks available. If those books are outdated or biased, the student's understanding will be limited. Similarly, AI models rely on preexisting data, which can constrain their ability to innovate or understand unfamiliar concepts. This reliance can sometimes make AI feel rigid, unable to step outside the patterns and information it has been fed.

One area where AI often stumbles is in contextual understanding, particularly when it comes to cultural references and idiomatic expressions. These nuances require a depth of understanding that goes beyond mere words. For instance, idioms like "break the ice" or "hit the hay" don't translate literally, and AI can struggle to interpret such phrases correctly. This limitation is akin to a traveler misunderstanding local customs due to language barriers. While AI can process vast amounts of text, it lacks the intuitive grasp of context that humans possess, making it challenging for AI to navigate conversations that rely heavily on cultural or contextual cues.

Real-world knowledge shortfalls also pose a challenge for AI. Unlike humans, AI lacks the ability to apply commonsense reasoning to scenarios. Consider a situation where you need to understand the implications of an unexpected event, like a sudden rainstorm during an outdoor event. Humans naturally adapt their plans, but AI, without real-world experience, may not recognize the practical steps needed to address such changes.

This gap in understanding can limit AI's effectiveness in dynamic environments where flexibility and real-world knowledge are crucial. It's a reminder that while AI can assist with information processing, it relies on human guidance for practical decision-making.

AI's ethical constraints are another significant area of concern. Bias and fairness are critical issues that continue to challenge AI developers. Bias can be introduced at various stages of AI model development, from the data used for training to the algorithms themselves. For example, if a dataset contains historical biases, those biases can be perpetuated in AI outputs. Efforts to enhance AI fairness and inclusivity are ongoing, with developers striving to create models that minimize bias and promote equitable outcomes. These initiatives are part of a broader movement to ensure AI serves all users fairly, regardless of background or demographic.

To navigate these challenges, it's important to engage with AI critically, recognizing its strengths while being mindful of its limitations. By understanding AI's dependency on data, its struggles with context, its lack of real-world experience, and ethical concerns, you can make informed decisions about how to incorporate AI into your work. As you continue to explore AI's potential, keep these factors in mind to maximize its benefits while mitigating potential pitfalls.

### Reflective Exercise: Evaluating AI Limitations

Take a moment to reflect on your recent interactions with AI. Consider a task where AI fell short and identify which limitation—data dependency, contextual understanding, real-world knowledge, or ethical constraints—was most evident. Think about how you might address this issue in future interactions. Would providing additional context help? Could you supplement AI with human insight for better outcomes? Use this reflection to guide your approach, ensuring you leverage AI effectively while staying aware of its constraints.

## 11.1 NAVIGATING AMBIGUOUS RESPONSES

When interacting with AI like ChatGPT, you might encounter responses that seem vague or unclear. This ambiguity often arises from the nature of the input provided. Imagine asking a friend for movie recommendations without mentioning your favorite genres or current mood. The suggestions might range widely, some hitting the mark while others miss entirely. Similarly, AI thrives on specificity and context. Without clear and detailed prompts, it struggles to pinpoint exactly what you want it to address. Context plays a significant role in shaping AI responses. A question like "What do you think?" without any prior information leaves the AI guessing at your intent. It can lead to generic or off-target replies, emphasizing the importance of clarity in communication.

To navigate and clarify these ambiguous outputs, consider refining your approach to questioning. Just as you might ask a friend to elaborate on a story they're telling, you can request additional information from the AI. Phrasing your questions with precision helps, too. Instead of broad queries, use specific language that guides the AI toward the exact information you're seeking. For example, instead of asking, "Tell me about climate change," you might specify, "Can you explain the impact of climate change on polar ice caps?" This not only narrows the focus but also helps the AI provide a more targeted and relevant response, reducing ambiguity.

Interpreting AI responses requires a bit of critical thinking. When faced with an output that seems ambiguous, take a moment to assess it critically. Consider what the AI might have missed or how the context could have influenced its response. By analyzing the AI's reply, you can often determine which elements need clarification. Contextualizing AI responses with external information can also be beneficial. Suppose you're working on a topic you know well. In that case, you can use your background knowledge to fill in the gaps, understanding the AI's output as part of a larger narrative. This approach helps make sense of responses that might initially seem lacking in detail or depth.

In critical or high-stakes situations, ambiguous AI outputs can pose significant challenges. Imagine relying on AI for crucial business decisions, only to receive a response that leaves more questions than answers. In these cases, it's vital to cross-verify AI outputs with human expertise. Consulting a knowledgeable colleague or expert can provide the insights needed to interpret and validate the AI's suggestions. Implementing fallback strategies is also crucial. These strategies involve having alternative plans or solutions in place should the AI's outputs not meet the required standards. By preparing for potential ambiguity, you ensure that critical decisions are not solely dependent on AI, maintaining a balance between technological assistance and human judgment.

## 11.2 STRATEGIES FOR RELIABLE AI OUTPUTS

To get the most reliable results from ChatGPT, refining how you structure your inputs is key. Imagine you're trying to explain a complex idea to a friend. If you use vague terms or leave out important context, the conversation might end up muddled. The same principle applies when interacting with AI. Using specific and detailed prompts helps the AI understand exactly what you're asking. This approach is like giving clear directions to someone unfamiliar with the area—they're more likely to reach the intended destination. Providing relevant background information further sharpens the AI's focus. Suppose you're asking for advice on a specific topic; adding context or previous attempts can guide the AI to deliver more useful and tailored responses. This structured approach in crafting prompts ensures that the AI can process your request accurately, minimizing the chances of receiving irrelevant or off-target outputs.

User feedback plays a pivotal role in enhancing AI performance. Think of it as a feedback loop where each interaction helps refine future responses. Engaging with ChatGPT and noting areas where it excels or falls short can guide improvements. For instance, if you notice the AI consistently misinterprets a certain type of question, providing feedback helps adjust its algorithms to improve accuracy. Implementing changes based on user input is akin to a coach refining an athlete's technique—each tweak enhances performance. Over time, this iterative process leads to a more responsive and reliable AI, catering to the specific needs of its users. By involving users in this way, AI systems become more attuned to human expectations and requirements, creating a more effective tool for everyday tasks.

Incorporating AI tools and plugins can further boost the reliability of AI outputs. These tools act like quality control checks, ensuring that AI responses meet high standards. Integrating AI moderation tools can help filter out inappropriate or irrelevant content, much like a seasoned editor refining an article before publication. These tools assess the AI's outputs, flagging areas that require improvement or adjustment. Meanwhile, leveraging quality assurance plugins ensures that the AI maintains consistency and accuracy across interactions. These plugins serve as an extra layer of scrutiny, verifying that the AI's responses align with the intended quality benchmarks. Together, these enhancements create a robust framework that supports reliable and precise AI outputs, allowing users to trust the system with confidence.

Developing verification protocols is another strategy to ensure the accuracy of AI responses. These protocols act as a safety net, catching discrepancies or errors before they affect decision-making. Establishing cross-checking procedures involves comparing AI outputs with established facts or data sources, much like a fact-checker verifying claims in a report. This step ensures that the AI's suggestions are grounded in reality and not based on flawed assumptions. Collaborating with subject matter experts can also enrich the verification process. Experts bring a depth of knowledge that AI might lack, providing insights that help refine and validate AI outputs. By combining AI's computational power with human expertise, you create a balanced approach that maximizes reliability while minimizing errors. This collaborative effort enhances trust in AI systems, making them valuable allies in achieving accurate and dependable outcomes.

## 11.3 MANAGING EXPECTATIONS WITH AI CAPABILITIES

When you first encounter AI, it might seem like a magical entity capable of solving all problems with the wave of a digital wand. However, setting realistic expectations is crucial for leveraging AI effectively. It's important to view AI as a supportive tool rather than a standalone solution. AI excels at processing vast amounts of information and providing insights, but it doesn't replace the nuanced decision-making that humans bring to the table. Recognizing the boundaries of AI functionality means understanding that while AI can assist with tasks like content generation or data analysis, it relies on human oversight to ensure these tasks align with broader goals and ethical standards. This perspective helps you appreciate AI as a valuable partner in your endeavors, enhancing your work without overshadowing the unique contributions that only human intuition and creativity can provide.

To bridge the gap between AI's potential and its limitations, educational resources play a vital role. Developing programs that foster AI literacy can empower users to engage with this technology confidently and responsibly. These programs can include workshops, online courses, or informational guides that demystify AI, explaining how it works and what it can realistically achieve. By providing clear, accessible information, users can better navigate their interactions with AI, making informed decisions about how to integrate it into their workflows. Informational guides, much like a user manual, can offer step-by-step advice on optimizing AI use, ensuring that users understand both the possibilities and the constraints of the technology. When users are equipped with this knowledge, they can set realistic expectations and leverage AI's strengths to their fullest potential.

The collaboration between humans and AI is where the real magic happens. By balancing AI use with human expertise, you create an environment where both strengths are utilized effectively. Collaborative workflows can enhance productivity by allowing AI to handle data-driven tasks while humans focus on strategic and creative aspects. For instance, in a marketing team, AI might analyze customer data to identify trends, while human team members use those insights to craft compelling campaigns. Delegating tasks based on AI and human strengths ensures that each aspect of a project benefits from the right expertise. This synergy not only improves efficiency but also fosters innovation, as humans and AI push each other to explore new possibilities and solutions.

Critically evaluating AI's impact on decision-making processes is another essential aspect of managing expectations. It's crucial to assess how AI influences strategic choices, ensuring that decisions are not solely based on data-driven insights but also consider human values and judgment. Implementing decision review mechanisms can help maintain this balance. These mechanisms involve regular evaluation of AI-driven decisions, allowing for adjustments and refinements based on human feedback and insights. By scrutinizing the outcomes of AI-assisted decisions, you can identify areas where human intervention might be necessary, ensuring that decisions align with organizational goals and ethical considerations. This ongoing evaluation helps refine AI's role within decision-making processes, ensuring that it remains a supportive tool rather than an unchecked influence.

As you navigate the complexities of AI integration, remember that managing expectations is a continuous process. It involves understanding AI's capabilities, educating users, fostering collaboration, and critically assessing its impact. By doing so, you create a dynamic environment where AI enhances human potential, driving innovation and efficiency while respecting the unique qualities that only humans can bring to the table. This balance is key to unlocking the full potential of AI, ensuring that it serves as a catalyst for growth and progress in your endeavors.

# CHAPTER TWELVE

## FUTURE OF AI AND CHATGPT

Imagine a bustling cityscape where vehicles glide effortlessly along streets, guided not by human hands but by sophisticated AI systems. These autonomous vehicles are no longer a distant vision; they are a reality shaping our daily commutes. AI-driven automation is increasingly redefining industries, streamlining processes, and enhancing efficiency in ways we are only beginning to comprehend. In transportation, AI not only pilots cars but also optimizes traffic flow to reduce congestion, cut emissions, and improve safety. The marvel of AI doesn't stop on the road. In healthcare, AI systems perform complex diagnostics with precision, transforming how we approach medical treatment. These systems analyze patient data to make accurate predictions about health outcomes, allowing doctors to tailor treatments more effectively. According to a study by the National Center for Biotechnology Information, AI in healthcare is enhancing patient care and streamlining administrative tasks. These examples illustrate just a fraction of AI's potential to revolutionize our world.

The evolution of machine learning models stands at the forefront of AI's transformative power. Neural networks, once simple in design, have grown into intricate architectures capable of processing vast amounts of data with remarkable speed and accuracy. These advances in machine learning enable AI to learn unsupervised, adapting to new information without the need for explicit programming. This capability allows AI to tackle complex problems, making decisions based on patterns it identifies in data. Such advancements not only improve existing applications but also pave the way for new innovations, from personalized marketing strategies to predictive analytics in finance. As machine learning models evolve, they unlock possibilities that were once confined to the realm of science fiction.

The integration of AI with the Internet of Things (IoT) marks another significant leap forward, promising to enhance connectivity and data utilization across various domains. In smart homes, AI-driven IoT devices analyze data from sensors to optimize energy use, improve security, and enhance overall living comfort. For instance, AI can adjust your home's heating and cooling systems based on your habits and weather forecasts, ensuring efficiency and comfort without manual intervention. A blog by Perpetio highlights how AI and IoT together create smarter, more responsive environments. In industrial settings, AI-powered IoT solutions monitor equipment performance, predict maintenance needs, and optimize production processes, reducing downtime and increasing efficiency. This fusion of AI and IoT not only makes life more convenient but also contributes to sustainable resource management, aligning with global efforts to reduce environmental impact.

The frontier of quantum computing presents a thrilling intersection where AI's capabilities could expand exponentially. Unlike traditional computers that use bits, quantum computers operate with qubits, which can process information at speeds unimaginable just a decade ago. This quantum leap in processing power holds the potential to revolutionize AI, enabling it to solve problems previously deemed unsolvable. In fields like cryptography, quantum-enhanced AI algorithms promise to enhance security measures, making data transmissions virtually unbreakable. Forbes reports on the potential for quantum AI to transform industries, from pharmaceuticals to weather forecasting, by providing insights that are both profound and precise. Quantum computing combined with AI opens new horizons, promising advancements that could redefine our technological landscape.

**Interactive Element: Visualizing AI's Future**

Imagine a world where AI seamlessly integrates into every aspect of our lives. Envision autonomous vehicles, smart homes, and quantum-enhanced AI. Sketch or write about how these technologies might interact in your daily life. Consider the benefits and challenges they might present. This exercise encourages you to think creatively about the future of AI and ChatGPT, visualizing its potential impact on society and your personal experiences.

## 12.1 PREPARING FOR AI'S EVOLVING ROLE IN SOCIETY

As AI continues to infiltrate various sectors, its impact on the workforce becomes increasingly evident. We're witnessing a shift where traditional roles evolve, and new ones emerge, necessitating a dynamic approach to skill development. The rise of AI ethics and governance roles is a testament to this change. These positions focus on ensuring AI systems operate within ethical boundaries, addressing concerns about bias and accountability. As AI permeates more aspects of our lives, its governance becomes paramount, requiring professionals who can navigate this complex terrain. Furthermore, reskilling programs are gaining traction, preparing workers to adapt to AI integration. These programs emphasize technical skills, such as data analysis and machine learning, alongside soft skills like problem-solving and collaboration. By fostering a workforce adept at leveraging AI, we can mitigate fears of job displacement and capitalize on the opportunities AI presents.

In the realm of public policy and governance, AI plays a crucial role in shaping future frameworks. Predictive policing, for instance, uses AI algorithms to analyze crime data, potentially preventing criminal activities before they occur. While this application promises enhanced security, it also raises ethical concerns regarding privacy and discrimination. Policymakers must therefore balance these benefits with the need for equitable practices. The development of policy frameworks for AI regulation is essential in ensuring these technologies serve the public good. Transparent and inclusive policymaking can address potential biases and foster trust in AI systems. By establishing clear guidelines, governments can harness AI's potential while safeguarding citizens' rights.

Education, too, is undergoing a transformation driven by AI. Adaptive learning environments tailor educational experiences to individual students, enhancing engagement and understanding. By analyzing student performance data, AI can identify strengths and weaknesses, providing personalized learning paths. This approach not only caters to diverse learning styles but also democratizes education by making quality resources accessible globally. Students in remote areas can now benefit from the same educational materials as those in well-funded schools, narrowing the educational divide. AI's role in education extends beyond personalized instruction; it also aids teachers by automating administrative tasks, allowing them to focus more on student interaction and less on paperwork.

However, the integration of AI into society is not without challenges. Social and ethical considerations must be at the forefront of AI development to ensure its benefits are widespread and equitable. Balancing privacy with innovation is a delicate task, requiring robust data protection measures without stifling technological progress. Equally important is ensuring that AI benefits are distributed fairly across communities, preventing the exacerbation of existing inequalities. This involves creating inclusive AI systems that consider diverse perspectives and needs. By engaging stakeholders from various backgrounds in the design and implementation of AI technologies, we can build systems that reflect our collective values and aspirations. These considerations are crucial as we navigate the complexities of AI's evolving role in society, striving for a future where technology enhances our lives while upholding our fundamental rights and freedoms.

## 12.2 PERSONAL GROWTH AND AI SYNERGIES

Imagine having a personal coach at your fingertips, ready to offer tailored advice at any time. AI-driven personal coaching apps are transforming how individuals approach personal development. These intelligent systems analyze your goals and behaviors, offering insights that help you grow. They track progress, suggest new strategies, and even provide motivational nudges to keep you on track. This personalized guidance allows you to focus on areas that matter most, whether it's improving productivity or developing new skills. Language learning, too, has evolved with AI tutors. These digital educators adapt lessons to fit your pace and learning style, making it easier to grasp complex concepts. With AI, learning a new language becomes less daunting, as interactive exercises and real-time feedback make the process engaging and effective.

Creativity and innovation are no longer confined to the realms of art and science; they are becoming integral to personal expression and problem-solving. AI-assisted creative writing tools, for example, offer an innovative way to explore storytelling. These tools suggest plot twists, refine dialogue, and even help develop characters, turning the writing process into a collaborative experience. Similarly, innovative idea generation platforms leverage AI to expand on your initial concepts, presenting new angles you might not have considered. They encourage out-of-the-box thinking and foster a creative mindset that can be applied across various fields. By using AI to enhance creativity, you open doors to new opportunities and insights that fuel personal and professional growth.

In the realm of health and wellness, AI technologies are making significant strides. AI-powered fitness tracking devices monitor your physical activity, providing detailed analytics that help you understand your health better. They offer personalized workout plans, track progress, and even remind you to stay active, turning fitness into a manageable and rewarding part of daily life. Meanwhile, mental health support via AI chatbots is becoming increasingly prevalent. These chatbots offer a non-judgmental space to discuss feelings and receive guidance, making mental health care more accessible. They provide coping strategies, relaxation techniques, and resources tailored to your needs, supporting your journey toward mental well-being. The integration of AI in health and wellness empowers individuals to take control of their health, fostering a proactive approach to self-care.

Lifelong learning is a concept that resonates strongly in our rapidly changing world, and AI serves as a catalyst for continuous personal development. Customized learning paths via AI platforms adapt to your learning style and pace, offering courses and materials that align with your interests and career goals. These platforms recommend content based on your progress, ensuring that your learning remains relevant and challenging. AI-driven career development resources further enhance this process by identifying skills gaps and suggesting ways to fill them. Whether through online courses, webinars, or interactive workshops, AI provides the tools you need to stay competitive and fulfilled in your professional life. By embracing AI as an ally in lifelong learning, you can navigate changes with confidence and curiosity, always ready to explore the next chapter of your personal and professional journey.

## 12.3 ENVISIONING THE NEXT ERA OF AI INNOVATION

In the landscapes of emerging markets, AI is not just a tool but a transformative force, reshaping industries and opening doors to new possibilities. In agriculture, AI-driven technology is revolutionizing how we grow and harvest crops. By using data from weather patterns and soil sensors, AI can predict the best times to plant and harvest, enhancing productivity and ensuring food security. This shift not only boosts yields but also reduces waste, making agriculture more sustainable. On the financial front, AI is a catalyst for inclusion, providing access to banking and financial services for populations previously underserved. Through AI-driven solutions, individuals in remote areas can now perform transactions, access credit, and manage savings via mobile devices. This democratization of finance empowers communities, fostering economic growth and stability.

Cross-industry collaborations are fueling AI innovation in unprecedented ways. When technology firms join forces with healthcare providers, for example, the results can be groundbreaking. Consider projects where AI is used to interpret medical images, enhancing diagnostic accuracy and speed. These collaborations leverage the strengths of each sector, creating solutions that neither could achieve alone. In research, cross-sectoral initiatives push the boundaries of what AI can do, combining data from varied sources to solve complex problems. These partnerships accelerate innovation, as insights from one field illuminate possibilities in another. By pooling resources and expertise, industries can address challenges more effectively, driving progress and creating value that extends beyond individual sectors.

The future of human-AI interaction promises a world where technology and creativity blend seamlessly. Imagine AI as a collaborator in creative industries, working alongside artists to generate new ideas and designs. This partnership allows creators to explore uncharted territories, using AI to experiment with forms and styles. Human-centered AI design principles focus on making these technologies intuitive and accessible, prioritizing user experience. This approach ensures that AI enhances rather than hinders human creativity, providing tools that are powerful yet easy to use. As AI becomes more integrated into creative processes, it offers a new canvas for innovation, where human imagination meets technological precision.

AI is also poised to tackle some of the most pressing global challenges, such as climate change and resource scarcity. In climate modeling, AI analyzes vast datasets to predict environmental changes, helping scientists and policymakers make informed decisions. These models can forecast weather patterns, assess the impact of human activities, and suggest mitigation strategies. AI's role in sustainable resource management is equally crucial. By optimizing the use of natural resources, AI helps industries reduce their environmental footprint. This technology analyzes consumption patterns and suggests ways to minimize waste, promoting efficiency and sustainability. Through AI analytics, businesses can achieve their goals while preserving the planet's resources for future generations.

As we move forward, the potential for AI to drive innovation and address global challenges is immense. Emerging markets, cross-industry collaborations, human-AI interaction, and sustainability efforts illustrate the transformative power of AI. By embracing these opportunities, we can shape a future where technology and humanity coexist harmoniously, enhancing our world in ways we are only beginning to imagine.

# CONCLUSION

As we reach the final pages of this journey, let's take a moment to reflect on the core objective of this book: to empower you to harness the power of ChatGPT in your daily tasks. Whether you're here to streamline your workflow, enhance creativity, or simply understand AI's transformative potential, the goal has been to provide you with practical tools and insights to achieve just that.

Each chapter has been a stepping stone, guiding you through the intricacies of AI and ChatGPT. We began by demystifying AI, comparing it to a digital assistant, and illustrating its role in various real-world applications. Delving into the mechanics of ChatGPT, we unraveled the complexities of neural networks and natural language processing, providing a foundation for understanding how this powerful tool works. We addressed common myths and misconceptions, ensuring a clear, unbiased perspective on AI's capabilities and limitations.

The journey continued with hands-on guidance for getting started with ChatGPT. From installation to overcoming tech anxiety, these chapters aimed to build your confidence and familiarity with the interface. We explored crafting effective prompts, a skill essential for maximizing AI interaction, and practical applications in daily tasks like email management and brainstorming. Each section was designed to be accessible, encouraging you to experiment and customize your experience.

Our discussion of overcoming challenges highlighted ways to manage information overload, find reliable sources, and navigate common pitfalls. These insights aimed to equip you with strategies for maintaining efficiency and accuracy in AI interactions. The chapters on ethical use emphasized the importance of responsible AI engagement, from ensuring data privacy to fostering bias-free interactions. These considerations are crucial as AI becomes more integrated into our lives.

Highlighting ChatGPT's unique contributions, we delved into its application in content creation, marketing, education, and freelancing. These examples showcased AI's versatility and potential to transform industries, offering tangible benefits across various fields. The exploration of human creativity with AI emphasized collaboration, underscoring the synergy possible when AI and human ingenuity come together.

Key takeaways for you include the importance of clear communication, ethical engagement, and continuous learning. By understanding these principles, you are better prepared to leverage AI effectively while being mindful of its limitations. The call to action here is simple: apply what you've learned. Whether it's drafting emails more efficiently or exploring creative ideas with AI, take these insights and use them to enhance your daily tasks.

As you venture beyond this book, I encourage you to embrace continuous learning and growth. The landscape of AI is ever-evolving, and staying informed will only enhance your ability to adapt and thrive. Join AI communities, participate in forums, and explore resources to deepen your understanding and skills. Your journey with AI is just beginning, and the possibilities are endless.

I want to express my gratitude for your engagement and curiosity throughout this book. Writing this has been a rewarding experience, and it's my hope that you find the insights and strategies as valuable as I intended. Remember, AI is a tool to enhance your capabilities, not replace them. Use it wisely, and let it inspire innovation and creativity in all your endeavors. Thank you for allowing me to guide you on this journey, and here's to your continued success and exploration in the world of AI.

**Title:** Keeping the Game Alive

Now that you're ready to dive into ChatGPT and Artificial Intelligence, it's your turn to help others get started too!

By leaving your honest review on Amazon, you'll show other readers where to find the information they need. You can help them discover the same excitement and knowledge you found in this book.

Thank you for your support! The world of ChatGPT and Artificial Intelligence grows when we share what we know, and you're a big part of that.

Click here to leave your review on Amazon.

# REFERENCES


- *ChatGPT for NLP Tasks https://indatalabs.com/blog/chatgpt-in-nlp*
- *Transformer Neural Networks: A Step-by-Step Breakdown https://builtin.com/artificial-intelligence/transformer-neural-network*

- *Debunking 5 artificial intelligence myths | Carlson School of https://carlsonschool.umn.edu/graduate/resources/debunking-5-artificial-intelligence-myths*

- *The Ethical Considerations of Artificial Intelligence https://www.captechu.edu/blog/ethical-considerations-of-artificial-intelligence*

- *How to Access ChatGPT on Any Device: A Comprehensive ... https://medium.com/@14prasadtambde/how-to-access-chatgpt-on-any-device-a-comprehensive-guide-7c4f52d6b748*

- *5 Ways to Help Students Overcome Technology Anxiety https://blog.cengage.com/5-ways-to-help-students-overcome-technology-anxiety/*

- *ChatGPT: 10 Resources for Learning and Implementation https://medium.com/@Ruben.Aster/chatgpt-10-resources-for-learning-and-implementation-pack-i-61f50c18502*

- *15+ Rules For Crafting Effective GPT Chat Prompts - Expandi https://expandi.io/blog/chat-gpt-rules/*

- *How To Write Effective Prompts For ChatGPT: 7 Essential ... https://www.forbes.com/sites/jodiecook/2023/06/26/how-to-write-effective-prompts-for-chatgpt-7-essential-steps-for-best-results/*

- *Common Pitfalls to Avoid in AI Prompt Writing - Medium https://medium.com/the-functional-technologist/common-pitfalls-to-avoid-in-ai-prompt-writing-610386fbfd3a*

- *Prompt Engineering for Effective Interaction with ChatGPT https://machinelearningmastery.com/prompt-engineering-for-effective-interaction-with-chatgpt/*

- *Prompt Optimization Techniques: A Guide to ... https://www.signitysolutions.com/tech-insights/prompt-optimization-guide*

- *Boosting Email Productivity with ChatGPT: How-To Guide https://www.linkedin.com/pulse/boosting-email-productivity-chatgpt-how-to-guide-james-rowe-1fhic*

- *How to Use AI for Brainstorming (Use Cases and Tools) https://clickup.com/blog/how-to-use-ai-for-brainstorming/*

- *How To Use ChatGPT For Research https://clickup.com/blog/how-to-use-chatgpt-for-research/*

- *19 Ways To Use ChatGPT To Boost Your Productivity - Persona https://www.personatalent.com/productivity/ways-to-use-chatgpt-to-boost-your-productivity/*

- *Overcoming information overload in the age of AI https://www.fastcompany.com/90953837/overcoming-information-overload-in-the-age-of-ai*

- *Evaluating AI-Generated Content - Using AI Tools in Your ... https://libguides.northwestern.edu/c.php?g=1309787&p=10620196*

- *ChatGPT: 5 Common Mistakes and How to Prevent Them https://kwan.com/blog/chatgpt-5-common-mistakes-and-how-to-prevent-them/*

- *Ethical concerns mount as AI takes bigger decision-making ... https://news.harvard.edu/gazette/story/2020/10/ethical-concerns-mount-as-ai-takes-bigger-decision-making-role/*

- *Prompt Engineering for Effective Interaction with ChatGPT https://machinelearningmastery.com/prompt-*

*engineering-for-effective-interaction-with-chatgpt/*

- *AI as a Co-Author: Exploring Collaborative Writing with ... https://medium.com/@Phannuman/ai-as-a-co-author-exploring-collaborative-writing-with-technology-42bd0bea789a*

- *ChatGPT in Market Research: A Definitive Guide https://www.voxpopme.com/learn/blog/chatgpt-in-market-research/*

- *Free ChatGPT Course with Certificate (2024) https://www.mygreatlearning.com/academy/learn-for-free/courses/chatgpt-for-beginners*

- *The Impact of the GDPR on Artificial Intelligence https://securiti.ai/impact-of-the-gdpr-on-artificial-intelligence/*

- *Mitigating Bias In AI and Ensuring Responsible AI https://leena.ai/blog/mitigating-bias-in-ai/*

- *Managing the Environmental Impact of AI https://www.captechu.edu/blog/environmental-impact-of-ai*

- *Building Trust In AI: The Case For Transparency https://www.forbes.com/sites/bernardmarr/2024*

*/05/03/building-trust-in-ai-the-case-for-transparency/*

- *Harnessing the power of ChatGPT to boost publisher ... https://blockthrough.com/blog/harnessing-the-power-of-chatgpt-to-boost-publisher-productivity/*
- *15 Best AI-Powered Content Generation and Strategy Tools https://www.marketingaiinstitute.com/blog/ai-tools-for-content-strategy*
- *How ChatGPT Provides Personalized Learning and ... https://www.linkedin.com/pulse/revolutionizing-education-how-chatgpt-provides-learning-walker-td7vc*
- *Top 10 AI Tools for Freelancers in 2024 https://ruul.io/blog/the-best-ai-tools-for-freelancers*
- *9 Best ChatGPT Use Cases for Creative Work https://www.superside.com/blog/chatgpt-use-cases*
- *Leveraging AI to Propel Interdisciplinary Research https://medium.com/@david.a.ragland/leveraging-ai-to-propel-interdisciplinary-research-49d67fd6a426*

- *AI-Generated Art: How Can Visual Artists Navigate This ... https://artmarketingnews.com/ai-generated-art/*
- *ChatGPT appropriation: A catalyst for creative performance ... https://www.sciencedirect.com/science/article/pii/S0160791X24001672*
- *Top 12 AI Communities to Join in 2023 https://2slash.ai/blog/top-ai-communities-2023/*
- *Top AI Resources for Learning: Courses and Online ... https://ai-pro.org/learn-ai/articles/the-top-ai-resources-for-learning-courses-and-online-platforms/*
- *Mentoring in the AI World https://chronus.com/blog/mentoring-in-the-ai-world*
- *The state of AI in 2023: Generative AI's breakout year https://www.mckinsey.com/capabilities/quantumblack/our-insights/the-state-of-ai-in-2023-generative-ais-breakout-year*
- *Understanding AI and Data Dependency https://www.dataideology.com/understanding-ai-and-data-dependency/*
- *Persistent Issues with GPT Model: Loss of Contextual ...*

*https://community.openai.com/t/persistent-issues-with-gpt-model-loss-of-contextual-understanding-and-continuity-in-conversations/600365*

- *Understanding Bias and Fairness in AI Systems https://towardsdatascience.com/understanding-bias-and-fairness-in-ai-systems-6f7fbfe267f3*

- *Prompt Engineering: Techniques, Applications, and Benefits https://www.spiceworks.com/tech/artificial-intelligence/articles/what-is-prompt-engineering/*

- *The potential for artificial intelligence in healthcare - PMC https://www.ncbi.nlm.nih.gov/pmc/articles/PMC6616181/*

- *Integrating IoT and AI for Effective Smart Home Automation https://perpet.io/blog/integrating-iot-and-ai-for-effective-smart-home-automation/*

- *Quantum Artificial Intelligence Is Closer Than You Think https://www.forbes.com/sites/jonathanreichental/2023/11/20/quantum-artificial-intelligence-is-closer-than-you-think/*

- *Ethics of Artificial Intelligence https://www.unesco.org/en/artificial-intelligence/recommendation-ethics*

Made in the USA
Monee, IL
04 November 2026

67380866R00105